GENETICS AND GENETIC ENGINEERING

ISSN 1546-6426

GENETICS AND GENETIC ENGINEERING

Barbara Wexler

INFORMATION PLUS® REFERENCE SERIES
Formerly Published by Information Plus, Wylie, Texas

THOMSON

GALE

Detroit • New York • San Francisco • New Haven, Conn. • Waterville, Maine • London

THOMSON

GALE

Genetics and Genetic Engineering
Barbara Wexler
Paula Kepos, Series Editor

Project Editors
Kathleen J. Edgar, John McCoy

Permissions
Jackie Jones, Jhanay Williams

Composition and Electronic Prepress
Evi Seoud

Manufacturing
Cynde Bishop

ISBN-13: 978-0-7876-5103-9 (set)
ISBN-10: 0-7876-5103-6 (set)
ISBN-13: 978-1-4144-0756-2
ISBN-10: 1-4144-0756-4
ISSN 1546-6426

This title is also available as an e-book.
ISBN-13: 978-1-4144-2948-9 (set), ISBN-10: 1-4144-2948-7 (set)
Contact your Gale Group sales representative for ordering information.

Printed in the United States of America
10 9 8 7 6 5 4 3 2 1

TABLE OF CONTENTS

PREFACE

Genetics and Genetic Engineering is part of the *Information Plus Reference Series*. The purpose of each volume of the series is to present the latest facts on a topic of pressing concern in modern American life. These topics include today's most controversial and most studied social issues: abortion, capital punishment, care of senior citizens, crime, the environment, health care, immigration, minorities, national security, social welfare, women, youth, and many more. Although written especially for the high school and undergraduate student, this series is an excellent resource for anyone in need of factual information on current affairs.

By presenting the facts, it is the Gale Group's intention to provide its readers with everything they need to reach an informed opinion on current issues. To that end, there is a particular emphasis in this series on the presentation of scientific studies, surveys, and statistics. These data are generally presented in the form of tables, charts, and other graphics placed within the text of each book. Every graphic is directly referred to and carefully explained in the text. The source of each graphic is presented within the graphic itself. The data used in these graphics are drawn from the most reputable and reliable sources, in particular from the various branches of the U.S. government and from major independent polling organizations. Every effort has been made to secure the most recent information available. The reader should bear in mind that many major studies take years to conduct and that additional years often pass before the data from these studies are made available to the public. Therefore, in many cases the most recent information available in 2007 dated from 2004 or 2005. Older statistics are sometimes presented as well if they are of particular interest and no more-recent information exists.

Although statistics are a major focus of the *Information Plus Reference Series*, they are by no means its only content. Each book also presents the widely held positions and important ideas that shape how the book's subject is discussed in the United States. These positions are explained in detail and, where possible, in the words of their proponents. Some of the other material to be found in these books includes: historical background; descriptions of major events related to the subject; relevant laws and court cases; and examples of how these issues play out in American life. Some books also feature primary documents or have pro and con debate sections giving the words and opinions of prominent Americans on both sides of a controversial topic. All material is presented in an even-handed and unbiased manner; the reader will never be encouraged to accept one view of an issue over another.

HOW TO USE THIS BOOK

Genetics and genetic engineering is a hotly debated topic in the United States today. Issues such as genetic testing to screen for disease susceptibility and the implications of test results; whether to permit cloning of human organs and the moral dilemma of "playing God"; and the advisability of raising genetically modified crops are discussed in coffee shops and classrooms across the nation. These topics and more are covered in this volume.

Genetics and Genetic Engineering consists of ten chapters and three appendixes. Each of the chapters is devoted to a particular aspect of genetics and genetic engineering in the United States. For a summary of the information covered in each chapter, please see the synopses provided in the Table of Contents at the front of the book. Chapters generally begin with an overview of the basic facts and background information on the chapter's topic, then proceed to examine subtopics of particular interest. For example, Chapter 9, Genetic Engineering and Biotechnology, begins by discussing agricultural applications of genetic engineering, then goes on to examine the creation of transgenic crops. After a discussion of whether genetically modified crops are helpful or harmful, the chapter delves into the U.S. biotechnology regulatory system. Americans' opinions regarding the consumption of genetically modified foods are also presented. Readers can find their way through a chapter by looking for the section and subsection headings, which are clearly set off from the text. They can also refer to the book's extensive Index if they already know what they are looking for.

Statistical Information

The tables and figures featured throughout *Genetics and Genetic Engineering* will be of particular use to the reader in learning about this issue. These tables and figures represent an extensive collection of the most recent and important statistics on genetics and genetic engineering and related issues—for example, graphics in the book cover the landmarks in the history of genetics, the diagram of DNA, public opinion on animal and human cloning, and public opinion on hazards of biotech foods. The Gale Group believes that making this information available to the reader is the most important way in which we fulfill the goal of this book: to help readers to understand the issues and controversies surrounding genetics and genetic engineering in the United States and to reach their own conclusions.

Each table or figure has a unique identifier appearing above it for ease of identification and reference. Titles for the tables and figures explain their purpose. At the end of each table or figure, the original source of the data is provided.

In order to help readers understand these often complicated statistics, all tables and figures are explained in the text. References in the text direct the reader to the relevant statistics. Furthermore, the contents of all tables and figures are fully indexed. Please see the opening section of the Index at the back of this volume for a description of how to find tables and figures within it.

Appendixes

In addition to the main body text and images, *Genetics and Genetic Engineering* has three appendixes. The first is the Important Names and Addresses directory. Here the reader will find contact information for a num-

ber of government and private organizations that can provide further information on genetics and genetic engineering. The second appendix is the Resources section, which can also assist the reader in conducting his or her own research. In this section the author and editors of *Genetics and Genetic Engineering* describe some of the sources that were most useful during the compilation of this book. The final appendix is the Index.

ADVISORY BOARD CONTRIBUTIONS

The staff of Information Plus would like to extend its heartfelt appreciation to the Information Plus Advisory Board. This dedicated group of media professionals provides feedback on the series on an ongoing basis. Their comments allow the editorial staff who work on the project to make the series better and more user-friendly. Our top priorities are to produce the highest-quality and most useful books possible, and the Advisory Board's contributions to this process are invaluable.

The members of the Information Plus Advisory Board are:

- Kathleen R. Bonn, Librarian, Newbury Park High School, Newbury Park, California

- Madelyn Garner, Librarian, San Jacinto College— North Campus, Houston, Texas

- Anne Oxenrider, Media Specialist, Dundee High School, Dundee, Michigan

- Charles R. Rodgers, Director of Libraries, Pasco-Hernando Community College, Dade City, Florida

- James N. Zitzelsberger, Library Media Department Chairman, Oshkosh West High School, Oshkosh, Wisconsin

COMMENTS AND SUGGESTIONS

The editors of the *Information Plus Reference Series* welcome your feedback on *Genetics and Genetic Engineering*. Please direct all correspondence to:

Editors
Information Plus Reference Series
27500 Drake Rd.
Farmington Hills, MI 48331-3535

CHAPTER 1
THE HISTORY OF GENETICS

Science seldom proceeds in the straightforward logical manner imagined by outsiders.

—James D. Watson, *The Double Helix: A Personal Account of the Discovery of the Structure of DNA* (1968)

Genetics is the biology of heredity, and geneticists are the scientists and researchers who study hereditary processes such as the inheritance of traits, distinctive characteristics, and diseases. Genetics considers the biochemical instructions that convey information from generation to generation.

Tremendous strides in science and technology have enabled geneticists to demonstrate that some genetic variation is related to disease and that the ability to vary genes improves the capacity of a species to survive changes in the environment. Although some of the most important advances in genetics research—such as deciphering the genetic code, isolating the genes that cause or predict susceptibility to certain diseases, and successfully cloning plants and animals—have occurred since the mid-twentieth century, the history of genetics study spans a period of about 150 years. As understanding of genetics progressed, scientific research became increasingly more specific. Genetics first considered populations, then individuals, then it advanced to explore the nature of inheritance at the molecular level.

EARLY BELIEFS ABOUT HEREDITY

From the earliest recorded history, ancient civilizations observed patterns in reproduction. Animals bore offspring of the same species, children resembled their parents, and plants gave rise to similar plants. Some of the earliest ideas about reproduction, heredity, and the transmission of information from parent to child were the particulate theories developed in ancient Greece during the fourth century B.C. These theories posited that information from each part of the parent had to be communicated to create the corresponding body part in the offspring. For example, the particulate theories held that information from the parent's heart, lungs, and limbs was transmitted directly from these body parts to create the offspring's heart, lungs, and limbs.

Particulate theories were attempts to explain observed similarities between parents and their children. One reason these theories were inaccurate was that they relied on observations unaided by the microscope. Microscopy—the use of or investigation with the microscope—and recognition of cells and microorganisms did not occur until the end of the seventeenth century, when the English naturalist Robert Hooke (1635–1703) first observed cells through a microscope.

Until that time (and even for some time after) heredity remained poorly understood. During the Renaissance (from about the fourteenth to the sixteenth centuries), preformationist theories proposed that the parent's body carried highly specialized reproductive cells that contained whole, preformed offspring. Preformationist theories insisted that when these specialized cells containing the offspring were placed in suitable environments, they would spontaneously grow into new organisms with traits similar to the parent organism.

The Greek philosopher Aristotle, who was such a keen observer of life that he is often referred to as the father of biology, noted that individuals sometimes resemble remote ancestors more closely than their immediate parents. He was a preformationist, positing that the male parent provided the miniature individual and the female provided the supportive environment in which it would grow. He also refuted the notion of a simple, direct transfer of body parts from parent to offspring by observing that animals and humans who had suffered mutilation or loss of body parts did not confer these losses to their offspring. Instead, he described a process that he termed *epigenesist*, in which the offspring is gradually generated from an undifferentiated mass by the addition of parts.

Of Aristotle's many contributions to biology, one of the most important was his conclusion that inheritance involved the potential of producing certain characteristics rather than the absolute production of the characteristics themselves. This thinking was closer to the scientific reality of inheritance than any philosophy set forth by his predecessors. However, because Aristotle was developing his theories before the advent of microscopy, he mistakenly presumed that inheritance was conveyed via the blood. Regardless, his enduring influence is evident in the language and thinking about heredity. Although blood is not the mode of transmission of heredity, people still refer to "blood relatives," "blood lines," and offspring as products of their own "flesh and blood."

One of the most important developments in the study of hereditary processes came in 1858, when Charles Darwin and Alfred Russel Wallace announced the theory of natural selection—the idea that members of a population who are better adapted to their environment will be the ones most likely to survive and pass their traits on to the next generation. Darwin published his theories in *On the Origin of Species by Means of Natural Selection* (1859). Darwin's work was not viewed favorably, especially by religious leaders who believed that it refuted the biblical interpretation of how life on Earth began. Even in the twenty-first century the idea that life evolves gradually through natural processes is not accepted by everyone, and the dispute over creationism and evolution continues.

CELL THEORY

In 1665, when Hooke used the microscope he had designed to examine a piece of cork, he saw a honeycomb pattern of rectangles that reminded him of cells, the chambers of monks in monasteries. His observations prompted scientists to speculate that living tissue as well as nonliving tissue was composed of cells. The French scientist René Dutrochet performed microscopic studies and concluded in 1824 that both plant and animal tissue was composed of cells.

In 1838 the German scientist Matthias Jakob Schleiden presented his theory that all plants were constructed of cells. The following year Theodor Schwann suggested that animals were also composed of cells. Both Schleiden and Schwann theorized that cells were all created using the same process. Although Schleiden's hypotheses about the process of cell formation were not entirely accurate, both he and Schwann are credited with developing cell theory. Describing cells as the basic units of life, they asserted that all living things are composed of cells, the simplest forms of life that can exist independently. Their pioneering work enabled other scientists to understand accurately how cells live, and the German pathologist

Rudolf Virchow launched theories of biogenesis when he posited in 1858 that cells reproduce themselves.

Improvements in microscopy and the increasing study of cytology—the formation, structure, and function of cells—enabled scientists to identify parts of the cell. Key cell components include the nucleus, which directs all cellular activities by controlling the synthesis of proteins, and the mitochondria, which are organelles (membrane-bound cell compartments) that catalyze reactions that produce energy for the cell. Figure 1.1 is a diagram of a typical animal cell that shows its component parts, including the contents of the nucleus, where chromosomes (which contain the genes) are located.

Germplasm Theory of Heredity

Studies of cellular components, processes, and functions produced insights that revealed the connection between cytology and inheritance. The German biologist August Weissmann studied medicine, biology, and zoology, and his contribution to genetics was an evolutionary theory known as the germplasm theory of heredity. Building on Darwin's idea that specific inherited characteristics are passed from one generation to the next, Weissmann asserted that the genetic code for each organism was contained in its germ cells (the cells that create sperm and eggs). The presence of genetic information in the germ cells explained how this information was conveyed, unchanged from one generation to the next.

In a series of essays about heredity published from 1889 to 1892, Weissmann observed that the amount of genetic material did not double when cells replicated, suggesting that there was some form of biological control of the chromosomes that occurred during the formation of the gametes (sperm and egg). His theory was essentially correct. Normal body growth is attributable to cell division, called mitosis, which produces cells that are genetically identical to the parent cells. The way to avoid giving offspring a double dose of heredity information is through a cell division that reduces the amount of the genetic material in the gametes by one-half. Weissmann called this process reduction division; it is now known as meiosis.

Weissmann was also the first scientist to successfully refute the members of the scientific community who believed that physical characteristics acquired through environmental exposure were passed from generation to generation. He conducted experiments in which he cut the tails off several consecutive generations of mice and observed that none of their offspring were born tailless.

A FARMER'S SON BECOMES THE FATHER OF GENETICS

Gregor Mendel was born on July 22, 1822, into a peasant family in what is now Hyncice, Czech Republic,

FIGURE 1.1

A typical animal cell

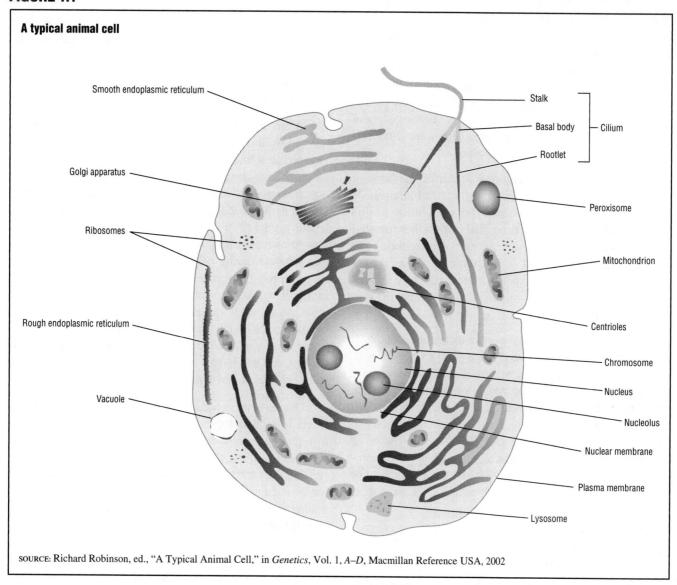

Smooth endoplasmic reticulum

Golgi apparatus

Ribosomes

Rough endoplasmic reticulum

Vacuole

Stalk

Basal body — Cilium

Rootlet

Peroxisome

Mitochondrion

Centrioles

Chromosome

Nucleus

Nucleolus

Nuclear membrane

Plasma membrane

Lysosome

SOURCE: Richard Robinson, ed., "A Typical Animal Cell," in *Genetics*, Vol. 1, *A–D*, Macmillan Reference USA, 2002

and spent much of his youth working in his family's orchards and gardens. At the age of twenty-one he entered the Abbey of St. Thomas, a Roman Catholic monastery, where he studied theology, philosophy, and science. His interest in botany (the scientific study of plants) and aptitude for natural science inspired his superiors to send him to the University of Vienna, where he studied to become a science teacher. However, Mendel was not destined to become an academic, despite his abiding interest in science and experimentation. In fact, the man who was eventually called the father of genetics never passed the qualifying examinations that would have enabled him to teach science at the highest academic level. Instead, he instructed students at a technical school. He also continued to study botany and conduct research at the monastery, and from 1868 until his death in 1884 he served as its abbot.

Between 1856 and 1863 Mendel conducted carefully designed experiments with nearly 30,000 pea plants he cultivated in the monastery garden. He chose to observe pea plants systematically because they had distinct, identifiable characteristics that could not be confused. Pea plants were also ideal subjects for his experiments because their reproductive organs were surrounded by petals and usually matured before the flower bloomed. As a result, the plants self-fertilized and each plant variety tended to be a pure breed. Mendel raised several generations of each type of plant to be certain that his plants were pure breeds. In this way, he confirmed that tall plants always produce tall offspring and plants with green seeds and leaves always produce offspring with green seeds and leaves.

His experiments were designed to test the inheritance of a specific trait from one generation to the next. For example, to test the inheritance of the characteristic of plant height, Mendel self-pollinated several short pea plants, and the seeds they produced grew into short plants. Similarly, self-pollinated tall plants and their resulting seeds, called the first or F1 generation, grew to be tall plants. These results seemed logical. When Mendel bred tall and short plants together and all their offspring in the F1 generation were tall, he concluded that the shortness trait had disappeared. However, when he self-pollinated the F1 generation, the offspring, called the F2 generation (second generation), contained both tall and short plants. After repeating this experiment many times, Mendel observed that in the F2 generation there were three tall plants for every short one—a 3:1 ratio.

Mendel's attention to rigorous scientific methods of observation, large sample size, and statistical analysis of the data he collected bolstered the credibility of his results. These experiments prompted him to theorize that characteristics, or traits, come in pairs—one from each parent—and that one trait will assume dominance over the other. The trait that appears more frequently is considered the stronger, or dominant, trait, whereas the one that appears less often is the recessive trait.

Focusing on plant height and other distinctive traits, such as the color of the pea pods, seed shape (smooth or wrinkled), and leaf color (green or yellow), enabled Mendel to record accurately and document the results of his plant breeding experiments. His observations about purebred plants and their consistent capacity to convey traits from one generation to the next represented a novel idea. The accepted belief of inheritance described a blending of traits, which, once combined, diluted or eliminated the original traits entirely. For example, it was believed that crossbreeding a tall and a short plant would produce a plant of medium height.

About the same time, Darwin was performing similar experiments using snapdragons, and his observations were comparable to those made by Mendel. Although Darwin and Mendel both explained the units of heredity and variations in species in their published works, it was Mendel who was later credited with developing the groundbreaking theories of heredity.

Mendel's Laws of Heredity

[T]he constant characters which appear in the several varieties of a group of plants may be obtained in all the associations which are possible according to the [mathematical] laws of combination, by means of repeated artificial fertilization.

—Gregor Mendel, "Versuche über Pflanzen-Hybriden" (1865)

From the results of his experiments, Mendel formulated and published three interrelated theories in the paper "Versuche über Pflanzen-Hybriden" (1865; translated into English as "Experiments in Plant Hybridization" in 1901). This work established the basic tenets of heredity:

• Two heredity factors exist for each characteristic or trait.

• Heredity factors are contained in equal numbers in the gametes.

• The gametes contain only one factor for each characteristic or trait.

• Gametes combine randomly, no matter which hereditary factors they carry.

• When gametes are formed, different hereditary factors sort independently.

When Mendel presented his paper, it was virtually ignored by the scientific community, which was otherwise engaged in a heated debate about Darwin's theory of evolution. Years later, well after Mendel's death in 1884, his observations and assumptions were revisited and became known as Mendel's laws of heredity. His first principle of heredity, the law of segregation, stated that hereditary units, now known as genes, are always paired and that genes in a pair separate during cell division, with the sperm and egg each receiving one gene of the pair. As a result, each gene in a pair will be present in half the sperm or egg cells. In other words, each gamete receives from a parent cell only one-half of the pair of genes it carries. Because two gametes (male and female) unite to reproduce and form a new cell, the new cell will have a unique pair of genes of its own, half from one parent and half from the other.

Diagrams of genetic traits conventionally use capital letters to represent the dominant traits and lowercase letters to represent recessive traits. Figure 1.2 uses this system to demonstrate Mendel's law of segregation. The pure red sweet pea and the pure white sweet pea each have two genes—RR for the red and rr for the white. The possible outcomes of this mating in the first generation are all hybrid (a combination of two different types) red plants (Rr)—plants that all have the same outward appearance (or phenotype) as the pure red parent but that also carry the white gene. As a result, when two of the hybrid first-generation plants are bred, there is a 50% chance that the resulting offspring will be hybrid red, a 25% chance that the offspring will be pure red, and a 25% chance that the offspring will be pure white.

Mendel also provided compelling evidence from his experiments for the law of independent assortment. This law established that each pair of genes is inherited independently of all other pairs. Figure 1.3 shows the chance distribution of any possible combination of traits. The F1 generation of tall flowering red and dwarf white sweet pea plants produced four tall hybrid red plants with the identical phenotype. However, each one has a combination of genetic information different from that of the original parent plants. The unique combination of genetic information is known as a genotype. The F2 generation, bred from two tall red hybrid flowers, produced four different phenotypes: tall with red flowers, tall with white flowers, dwarf with red flowers, and dwarf with white flowers. Both Figure 1.2 and Figure 1.3 demonstrate that recessive traits that disappear in the F1 generation may reappear in future generations in definite, predictable percentages.

FIGURE 1.2

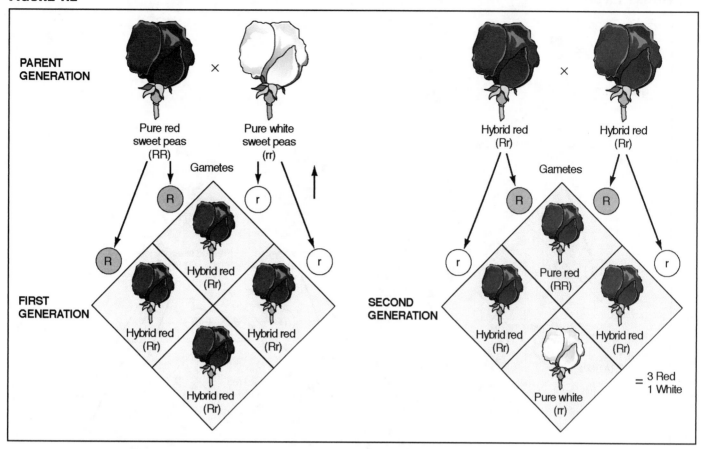

Mendel's law of segregation. *Hans & Cassidy, Thomson Gale.*

The law of dominance, the third tenet of inheritance identified by Mendel, asserts that heredity factors (genes) act together as pairs. When a cross occurs between organisms that are pure for contrasting traits, only one trait, the dominant one, appears in the hybrid offspring. In Figure 1.2 all the first-generation offspring are red—an identical phenotype to the parent plant—though they also carry the recessive white gene.

Mendel's contributions to the understanding of heredity were not acknowledged during his lifetime. When his efforts to reproduce the findings from his pea plant studies using hawkweed plants and honeybees did not prove successful, Mendel was dispirited. He set aside his botany research and returned to monastic life until his death. It was not until the early twentieth century, nearly forty years after he published his findings, that the scientific community resurrected Mendel's work and affirmed the importance of his ideas.

GENETICS AT THE DAWN OF THE TWENTIETH CENTURY

During the years following Mendel's work, understanding of cell division and fertilization increased, as did insight into the component parts of cells known as subcellular structures. For example, in 1869 the Swiss biochemist Johann Friedrich Miescher looked at pus he

had scraped from the dressings of soldiers wounded in the Crimean War (1853–56). In the white blood cells from the pus, and later in salmon sperm, he identified a substance he called nuclein. In 1874 Miescher separated nuclein into a protein and an acid, and it was renamed nucleic acid. He proposed that it was the "chemical agent of fertilization."

In 1900 three scientists—Karl Erich Correns, Hugo Marie de Vries, and Erich Tschermak von Seysenegg—independently rediscovered and verified Mendel's principles of heredity, and Mendel's contributions to modern genetics were finally acknowledged. In 1902 Sir Archibald E. Garrod, an English physician and chemist, applied Mendel's principles and identified the first human disease attributable to genetic causes, which he termed *inborn errors of metabolism*. The disease was alkaptonuria, a condition in which an abnormal buildup of an acid (homogentisic acid or alkapton) accumulates.

Seven years later, Garrod published *Inborn Errors of Metabolism*, a textbook describing various disorders that he believed were caused by these inborn metabolic errors. These included albinism (a pigment disorder in which affected individuals have abnormally pale skin, hair, and eyes) and porphyria (a group of disorders resulting from abnormalities in the production of heme, a vitally

FIGURE 1.3

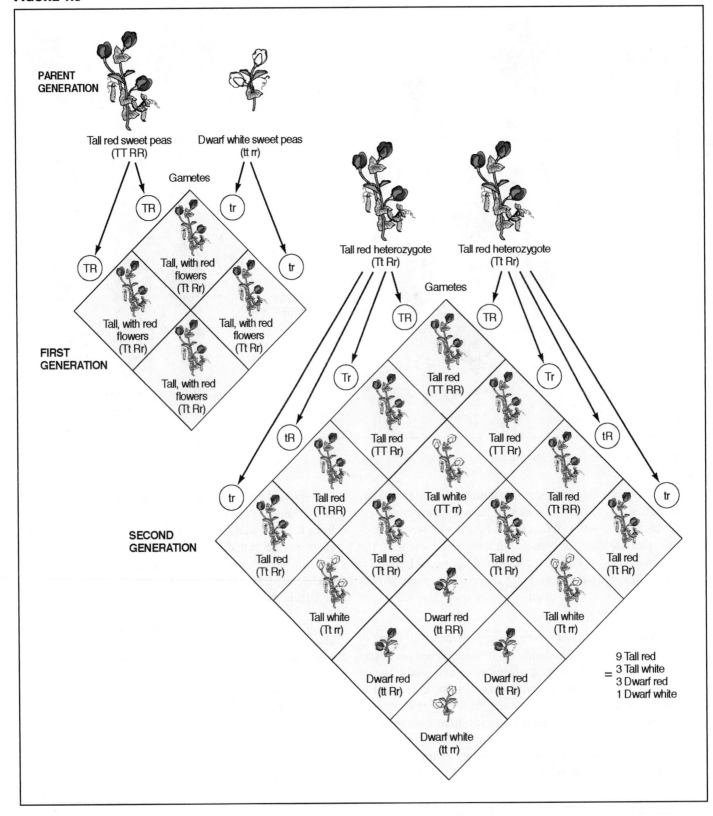

Mendel's law of independent assortment. *Hans & Cassidy, Thomson Gale.*

important substance that carries oxygen in the blood, bone, liver, and other tissues). Garrod's was the first effort to distinguish diseases caused by bacteria from those attributable to genetically programmed enzyme deficiencies that interfered with normal metabolism.

In 1905 the English geneticist William Bateson coined the term *genetics*, along with other descriptive terms used in modern genetics, including *allele* (a particular form of a gene), *zygote* (a fertilized egg), *homozygote* (an individual with genetic information that contains two identical forms of a gene), and *heterozygote* (an individual with two different forms of a particular gene). Arguably, his more important contributions to the progress of genetics were his translations of Mendel's work from German to English and his vigorous endorsement and promotion of Mendel's principles.

In 1908 the English mathematician Godfrey Harold Hardy and the German physician Wilhelm Weinberg independently developed a mathematical formula that describes the actions of genes in populations. Their assumptions that algebraic formulas could be used to analyze the occurrence of, and reasons for, genetic variation became known as the Hardy-Weinberg equilibrium. It advanced the application of Mendel's laws of heredity from individuals to populations, and by applying Mendelian genetics to Darwin's theory of evolution, it improved geneticists' understanding of the origin of mutations and how natural selection gives rise to hereditary adaptations. The Hardy-Weinberg equilibrium enables present-day geneticists to determine whether evolution is occurring in populations.

Chromosome Theory of Inheritance

Bateson is often cited for having said, "Treasure your exceptions." I believe Sturtevant's admonition would be, "Analyze your exceptions."

—E. B. Lewis, "Remembering Sturtevant," *Genetics*, 1995

The American geneticist and biologist Walter S. Sutton conducted studies using grasshoppers (*Brachystola magna*) he collected at his family's farm in Kansas. Sutton was strongly influenced by reading William Bateson's work and sought to clarify the role of the chromosomes in sexual reproduction. The results of his research, published in 1902, demonstrated that chromosomes exist in pairs that are structurally similar and proved that sperm and egg cells each have one pair of chromosomes. Sutton's work advanced genetics by identifying the relationship between Mendel's laws of heredity and the role of the chromosome in meiosis.

Along with Bateson, the American geneticist Thomas Hunt Morgan is often referred to as the father of classical genetics. In 1907 Morgan performed laboratory research using the fruit fly *Drosophila melanogaster*. He chose to study fruit flies because they bred quickly, had distinctive characteristics, and had just four chromosomes. The aim of his research was to replicate the genetic variation de Vries had reported from his experiments with plants and animals.

Working in a laboratory they called the "Fly Room," Morgan and his students Calvin B. Bridges, Hermann Joseph Muller, and Alfred H. Sturtevant conducted research that unequivocally confirmed the findings and conclusions of Mendel, Bateson, and Sutton. Breeding both white- and red-eyed fruit flies, they demonstrated that all the offspring were red eyed, indicating that the white-eye gene was recessive and the red-eye gene was dominant. The offspring carried the white-eye gene but it did not appear in the first generation. When, however, the F1 offspring were crossbred, the ratio of red-eyed to white-eyed flies was 3:1 in the next generation (F2). (A similar pattern is shown for red and white flowers in Figure 1.2.)

The investigators also observed that all the white-eyed flies were male, prompting them to investigate sex chromosomes and hypothesize about sex-linked inheritance. The synthesis of their research with earlier work produced the chromosomal theory of inheritance, the premise that genes are the fundamental units of heredity and are found in the chromosomes. It also confirmed that specific genes are found on specific chromosomes, that traits found on the same chromosome are not always inherited together, and that genes are actual physical objects. In 1915 the four researchers published *The Mechanism of Mendelian Heredity*, which detailed the results of their research, conclusions, and directions for future research.

In *The Theory of the Gene* (1926), Morgan asserted that the ability to quantify or number genes enables researchers to predict accurately the distribution of specific traits and characteristics. He contended that the mathematical principles governing genetics qualify it as science:

That the characters of the individual are referable to paired elements (genes) in the germinal matter that are held together in a definite number of linkage groups. . . . The members of each pair of genes separate when germ cells mature. . . . Each germ cell comes to contain only one set. . . . These principles . . . enable us to handle problems of genetics in a strictly numerical basis, and allow us to predict . . . what will occur. . . . In these respects the theory [of the gene] fulfills the requirements of a scientific theory in the fullest sense.

In 1933 Morgan was awarded the Nobel Prize in Physiology or Medicine for his groundbreaking contributions to the understanding of inheritance. Muller also became a distinguished geneticist, and after pursuing research on flies to determine if he could induce genetic changes using radiation, he turned his attention to studies of twins to gain a better understanding of human genetics. In 1946 he was awarded a Nobel Prize for his research on mutations, the source of all genetic variation.

Bridges eventually discovered the first chromosomal deficiency as well as chromosomal duplication in fruit flies. He served in various academic capacities at Columbia University, the Carnegie Institution, and the California Institute of Technology and was a member of the National Academy of Sciences and a fellow of the American Association for the Advancement of Science.

Sturtevant was awarded the National Medal of Science in 1968. His most notable contribution to genetics was the detailed outline and instruction he provided about gene mapping—the process of determining the linear sequence of genes in genetic material. In 1913 he began construction of a chromosome map of the fruit fly that was completed in 1951. Because of his work in gene mapping, he is often referred to as the father of the Human Genome Project, the comprehensive map of humanity's 20,000 to 25,000 genes. His book, *A History of Genetics* (1965), recounts the ideas, events, scientists, and philosophies that shaped the development of genetics.

CLASSICAL GENETICS

Another American geneticist awarded a Nobel Prize was Barbara McClintock, who described key methods of exchange of genetic information. Performing chromosomal studies of maize in the botany department at Cornell University, she observed colored kernels on an ear of corn that should have been clear. McClintock hypothesized that the genetic information that normally would have been conveyed to repress color had somehow been lost. She explained this loss by seeking and ultimately producing cytological proof of jumping genes, which could be released from their original position and inserted, or transposed, into a new position. This genetic phenomenon of chromosomes exchanging pieces became known as crossing over, or recombination.

With another pioneering female researcher, Harriet Creighton, McClintock published a series of research studies, including a 1931 paper that offered tangible evidence that genetic information crossed over during the early stages of meiosis (cell division). Along with the 1983 Nobel Prize in Physiology or Medicine, McClintock received the prestigious Albert Lasker Basic Medical Research Award in 1981, making her the most celebrated female geneticist in history.

During the same period, the English medical microbiologist Frederick Griffith performed experiments with *Streptococcus pneumoniae*, demonstrating that the ability to cause deadly pneumonia in mice could be transferred from one strain of bacteria to another. Griffith observed that the hereditary ability of bacteria to cause pneumonia could be altered by a "transforming principle." Although Griffith mistakenly believed the transforming factor was a protein, his observation offered the first tangible evidence linking deoxyribonucleic acid (DNA, the molecule that carries the genetic code) to heredity in cells. His experiment provided a framework for researching the biochemical basis of heredity in bacteria. In 1944 the Canadian-American immunologist Oswald T. Avery, along with the Canadian-American microbiologist Colin Munro Macleod and the American bacteriologist Maclyn McCarty, performed studies demonstrating that Griffith's transforming factor was DNA rather than simply a protein. Among the

FIGURE 1.4

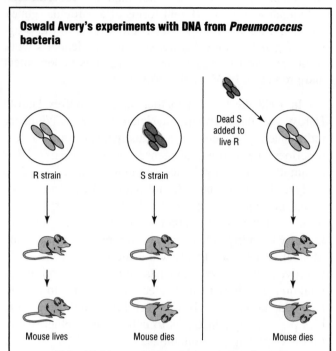

Oswald Avery's experiments with DNA from *Pneumococcus* bacteria

Dead S added to live R

R strain

S strain

Mouse lives

Mouse dies

Mouse dies

SOURCE: Richard Robinson, ed., "Oswald Avery's Experiments Showed that DNA from Dead S Strain of *Pneumococcus* Bacteria Could Transform a Harmless Strain into a Deadly Strain," in *Genetics*, Vol. 1, *A–D*, Macmillan Reference USA, 2002

experiments Avery, Macleod, and McCarty performed was one similar to Griffith's, which confirmed that DNA from one strain of bacteria could transform a harmless strain of bacteria into a deadly strain. (See Figure 1.4.) Their findings gave credence to the premise that DNA was the molecular basis for genetic information.

Nearly half of the twentieth century was devoted to classical genetics research and the development of increasingly detailed and accurate descriptions of genes and their transmission. In 1929 the Russian-American organic chemist Phoebus A. Levene isolated and discovered the structure of the individual units of DNA. Called nucleotides, the molecular building blocks of DNA are composed of deoxyribose (a sugar molecule), a phosphate molecule, and four types of nucleic acid bases. (See Figure 1.5.) Also in 1929 Theophilus Shickel Painter, an American cytologist, made the first estimate of the number of human chromosomes. His count of forty-eight was off by only two—twenty-five years later researchers were able to stain and view human chromosomes microscopically to determine that they number forty-six. Analysis of chromosome number and structure would become pivotal to medical diagnosis of diseases and disorders associated with altered chromosomal numbers or structure.

Another milestone in the first half of the twentieth century was the determination by the American chemist Linus C. Pauling that sickle-cell anemia (the presence of

FIGURE 1.5

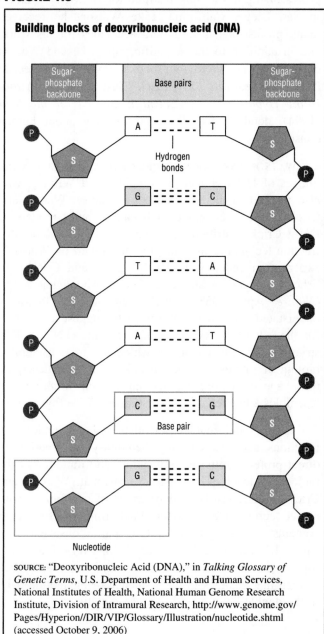

Building blocks of deoxyribonucleic acid (DNA)

Sugar-phosphate backbone | Base pairs | Sugar-phosphate backbone

A ----- T

Hydrogen bonds

G ===== C

T ----- A

A ----- T

C ===== G

Base pair

G ===== C

Nucleotide

SOURCE: "Deoxyribonucleic Acid (DNA)," in *Talking Glossary of Genetic Terms*, U.S. Department of Health and Human Services, National Institutes of Health, National Human Genome Research Institute, Division of Intramural Research, http://www.genome.gov/Pages/Hyperion//DIR/VIP/Glossary/Illustration/nucleotide.shtml (accessed October 9, 2006)

recounted as the "Waring blender experiment," the investigators dislodged virus particles that infect bacteria by spinning them in a blender and found that the viral DNA, and not the viral protein, that remains inside the bacteria directed the growth and multiplication of new viruses.

MODERN GENETICS EMERGES

The period of classical genetics focused on refining and improving the structural understanding of DNA. In contrast, modern genetics seeks to understand the processes of heredity and how genes work.

Many historians consider 1953—the year that the American geneticist James D. Watson and the English biophysicist Francis H. C. Crick famously described the structure of DNA—as the birth of modern genetics. It is important, however, to remember that Watson and Crick's historic accomplishment was not the discovery of DNA—Miescher had identified nucleic acid in cells nearly a century earlier. Similarly, even though Watson and Crick earned recognition and public acclaim for their landmark research, it would not have been possible without the efforts of their predecessors and colleagues such as Maurice H. F. Wilkins and Rosalind Elsie Franklin. Wilkins and Franklin were the molecular biologists who in 1951 obtained sharp X-ray diffraction photographs of DNA crystals, revealing a regular, repeating pattern of molecular building blocks that correspond to the components of DNA. (Wilkins shared the Nobel Prize with Watson and Crick, but Franklin was ineligible to share the prize because she died in 1958, four years before it was awarded.)

Another pioneer in biochemistry, the Austrian-American Erwin Chargaff, also provided information about DNA that paved the way for Watson and Crick. Chargaff suggested that DNA contained equal amounts of the four nucleotides: the nitrogenous (containing nitrogen, a nonmetallic element that constitutes almost four-fifths of the air by volume) bases adenine (A) and thymine (T), and guanine (G) and cytosine (C). In DNA there is always one A for each T, and one G for each C. This relationship became known as base pairing or Chargaff's rules, which also includes the observation that the ratio of AT to GC varies from species to species but remains consistent across different cell types within each species. (See Figure 1.6.)

Watson and Crick

James Watson is an American geneticist known for his willingness to grapple with big scientific challenges and his expansive view of science. In *The Double Helix: A Personal Account of the Discovery of the Structure of DNA* (1968), he chronicles his collaboration with Francis Crick to create an accurate model of DNA. He credits his

oxygen-deficient, abnormal red blood cells that cause affected individuals to suffer from pain and leg ulcers) was caused by the change in a single amino acid (a building block of protein) of hemoglobin (the oxygen-bearing, iron-containing protein in red blood cells). Pauling's work paved the way for research showing that genetic information is used by cells to direct the synthesis of protein and that mutation (a change in genetic information) can directly cause a change in a protein. This explains hereditary genetic disorders such as sickle-cell anemia.

From 1950 to 1952 the American geneticists Martha Cowles Chase and Alfred Day Hershey conducted experiments that provided definitive proof that DNA was genetic material. In research that would be widely

FIGURE 1.6

$$A + G = C + T$$

and

$$A = T \quad G = C$$

Chargaff's rules. *Argosy Publishing, Thomson Gale.*

inclination to take intellectual risks and venture into uncharted territory as his motivation for this ambitious undertaking.

Watson was just twenty-five years old when he announced the triumph that was hailed as one of the greatest scientific achievements of the twentieth century. Following this remarkable accomplishment, Watson served on the faculty of Harvard University for two decades and assumed the directorship and then the presidency of the Cold Spring Harbor Laboratory in Long Island, New York. From 1989 to 1992 he headed the Human Genome Project of the National Institutes of Health (NIH), the effort to sequence (or discover the order of) the entire human genome.

Crick was an English scientist who had studied physics before turning his attention to biochemistry and biophysics. Crick became interested in discovering the structure of DNA, and when in 1951 Watson came to work at the Cavendish Laboratory in Cambridge, England, the two scientists decided to work together to unravel the structure of DNA.

After his landmark accomplishment with Watson, Crick continued to study the relationship between DNA and genetic coding. He is credited with predicting the ways in which proteins are created and formed, a process known as protein synthesis. During the mid-1970s Crick turned his attention to the study of brain functions, including vision and consciousness, and assumed a

professorship at the Salk Institute for Biological Studies in San Diego, California. Like Watson, he received many professional awards and accolades for his work, and in addition to the scientific papers he and Watson coauthored, he wrote four books. Published a decade before his death in 2004, Crick's last book, *The Astonishing Hypothesis: The Scientific Search for the Soul* (1994), detailed his ideas and insights about human consciousness.

WATSON AND CRICK MODEL OF DNA. Using the X-ray images of DNA created by Franklin and Wilkins, who also worked in the Cavendish Laboratory, Watson and Crick worked out and then began to build models of DNA. Crick contributed his understanding of X-ray diffraction techniques and imaging and relied on Watson's expertise in genetics. In 1953 Watson and Crick published the paper "A Structure of Deoxyribonucleic Acid" (*Nature*, April 1953), which contained the famously understated first lines, "We wish to suggest a structure for the salt of deoxyribose nucleic acid (D.N.A.). This structure has novel features which are of considerable biological interest." Watson and Crick then described the shape of a double helix, an elegant structure that resembles a latticework spiral staircase. (See Figure 1.7.)

Their model enabled scientists to better understand functions such as carrying hereditary information to direct protein synthesis, replication, and mutation at the molecular level. The three-dimensional Watson and Crick model consists of two strings of nucleotides connected across like a ladder. Each rung of the ladder contains an A-T pair or a G-C pair, consistent with Chargaff's rule that there is an A for every T and a G for every C in DNA. (See Figure 1.7.) Watson and Crick posited that changes in the sequence of nucleotide pairs in the double helix would produce mutations.

MILESTONES IN MODERN GENETICS

During the second half of the twentieth century, geneticists and other researchers made remarkable strides. In 1956 Vernon M. Ingram, who would soon be recognized as the father of molecular medicine, identified the single base difference between normal and sickle-cell hemoglobin. The implications of his finding that the mutation of a single letter in the DNA genetic code was sufficient to cause a hereditary medical disorder were far reaching. This greater insight into the mechanisms of sickle-cell disease suggested directions for research into prevention and treatment. It prompted research that uncovered other diseases with similar causes, such as hemophilia (an inherited blood disease associated with insufficient clotting factors and excessive bleeding) and cystic fibrosis (an inherited disease of the mucous glands that produces problems associated with the lungs and

FIGURE 1.7

Canonical B–DNA double helix models

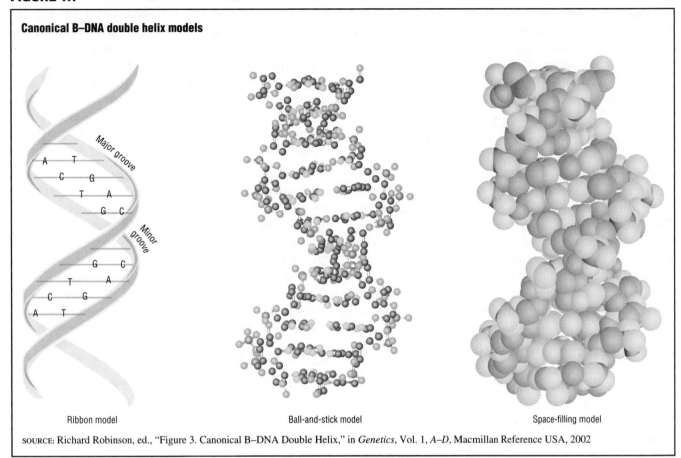

Ribbon model · Ball-and-stick model · Space-filling model

SOURCE: Richard Robinson, ed., "Figure 3. Canonical B–DNA Double Helix," in *Genetics*, Vol. 1, *A–D*, Macmillan Reference USA, 2002

pancreas). Just three years later, the first human chromosome abnormality was identified: people with Down syndrome were found to have an extra chromosome, demonstrating that it is a genetic disorder that may be diagnosed by direct examination of the chromosomes.

Ingram's work has been the foundation for current research to map genetic variations that affect human health. For example, in 1989, more than thirty years after Ingram's initial work, the gene for cystic fibrosis was identified and a genetic test for the gene mutation was developed.

Using radioactive labeling to track each strand of the DNA in bacteria, the American microbiologist Matthew S. Meselson and his colleague Franklin W. Stahl demonstrated with an experiment in 1958 that the replication of DNA in bacteria is semiconservative. Semiconservative replication occurs as the double helix unwinds at several points and knits a new strand along each of the old strands. Meselson and Stahl's experiment revealed that one strand remained intact and combined with a newly synthesized strand when DNA replicated, precisely as Watson and Crick's model predicted. In other words, each of the two new molecules created contains one of the two parent strands and one new strand.

In the early 1960s Crick, the American biochemist Marshall Nirenberg, the Russian-born American physicist George Gamow, and other researchers performed experiments that detected a direct relationship between DNA nucleotide sequences and the sequence of the amino acid building blocks of proteins. They determined that the four nucleotide letters (A, T, C, and G) may be combined into sixty-four different triplets. The triplets are code for instructions that determine the amino acid structure of proteins. Ribosomes are cellular organelles (membrane-bound cell compartments) that interpret a sequence of genetic code three letters at a time and link together amino acid building blocks of proteins specified by the triplets to construct a specific protein. The sixty-four triplets of nucleotides that can be coded in the DNA—which are copied during cell division, infrequently mutate and are read by the cell to direct protein synthesis—make up the universal genetic code for all cells and viruses.

Origins of Genetic Engineering

The late 1960s and early 1970s were marked by research that would lay the groundwork for modern genetic engineering technology. In 1966 DNA was found to be present not only in chromosomes but also in the mitochondria. The first single gene was isolated in 1969, and the following year the first artificial gene was created. In 1972 the American biochemist Paul Berg developed a technique to splice DNA fragments from different organisms and created the first recombinant DNA, or DNA molecules formed by combining segments of DNA, usually from different types of organisms. In 1980 Berg was awarded the Nobel Prize in Chemistry for this achievement, now referred to as recombinant DNA technology.

In 1976 an artificial gene inserted into a bacterium functioned normally. The following year DNA from a virus was fully decoded, and three researchers, working independently, developed methods to sequence DNA—in other words, to determine how the building blocks of DNA (the nucleotides A, C, G, and T) are ordered along the DNA strand. In 1978 bacteria were engineered to produce insulin, a pancreatic hormone that regulates carbohydrate metabolism by controlling blood glucose levels. Just four years later, the Eli Lilly pharmaceutical company marketed the first genetically engineered drug: a type of human insulin grown in genetically modified bacteria.

In 1980 the U.S. Supreme Court decision in *Diamond v. Chakrabarty* (404 U.S. 303) permitted patents for genetically modified organisms; the first one was awarded to the General Electric Company for bacteria to assist in clearing oil spills. The following year, a gene was transferred from one animal species to another. In 1983 the first artificial chromosome was created and the marker—the usually dominant gene or trait that serves to identify genes or traits linked with it—for Huntington's disease (an inherited disease that affects the functioning of both the body and brain) was identified; in 1993 the disease gene was identified.

In 1984 the observation that some nonfunctioning DNA is different in each individual launched research to refine tools and techniques developed by Sir Alec John Jeffreys at the University of Leicester in England that perform "genetic fingerprinting." Initially, the technique was used to determine the paternity of children, but it rapidly gained acceptance among forensic medicine specialists, who are often called on to assist in the investigation of crimes and interpret medicolegal issues.

The 1985 invention of the polymerase chain reaction (PCR), which amplifies (or produces many copies of) DNA, enabled geneticists, medical researchers, and forensic specialists to analyze and manipulate DNA from the smallest samples. PCR allowed biochemical analysis of even trace amounts of DNA. In *A Short History of Genetics and Genetic Engineering* (2003, http://www.dna50.com/dna 50.swf), Ricki Lewis and Bernard Possidente describe American biochemist Kary B. Mullis's development of PCR as the "genetic equivalent of a printing press," with the potential to revolutionize genetics in the same way that the printing press had revolutionized mass communications.

Five years later, in 1990, the first gene therapy was administered. Gene therapy introduces or alters genetic material to compensate for a genetic mistake that causes disease. The patient was a four-year-old girl with the inherited immune deficiency disorder adenine deaminase deficiency. If left untreated, the deficiency is fatal. Given along with conventional medical therapy, the gene therapy treatment was considered effective. The 1999 death of another gene therapy patient, as a result of an immune reaction to the treatment, tempered enthusiasm for gene therapy and prompted medical researchers to reconsider its safety and effectiveness.

Cloning—the production of genetically identical organisms—was performed first with carrots. A cell from the root of a carrot plant was used to generate a new plant. By the early 1950s scientists had cloned tadpoles, and during the 1970s attempts were underway to clone mice, cows, and sheep. These clones were created using embryos and many did not produce healthy offspring, offspring with normal life spans, or offspring with the ability to reproduce. In 1993 researchers at George Washington University in Washington, D.C., cloned nearly fifty human embryos, but their experiment was terminated after just six days.

In 1996 the English embryologist Ian Wilmut and his colleagues at the Roslin Institute in Scotland successfully cloned the first adult mammal that was able to reproduce. Dolly the cloned sheep, named for the country singer Dolly Parton, focused public attention on the practical and ethical considerations of cloning.

Human Genome Project and More

The term *genetics* refers to the study of a single gene at a time, whereas *genomics* is the study of all genetic information contained in a cell. The Human Genome Project (HGP) set as one of its goals the determination of the entire nucleotide sequence of the more than three billion bases of DNA contained in the nucleus of a human cell. Initial discussions about the feasibility and value of conducting the HGP began in 1986. The following year the first automated DNA sequencer was produced commercially. Automated sequencing, which enabled researchers to decode millions, as opposed to thousands, of letters of genetic code per day, was a pivotal technological advance for the HGP, which began in 1987 under the auspices of the U.S. Department of Energy (DOE).

In 1988 the HGP was relocated to the National Institutes of Health (NIH), and Watson was recruited to direct the project. The following year the NIH opened the National Center for Human Genome Research, and a committee composed of professionals from the NIH and DOE was named to consider ethical, social, and legal issues that might arise from the project. In 1990 the project began in earnest, with work on preliminary genetic maps of the human genome and four other organisms believed to share many genes with humans.

During the early 1990s several new technologies were developed that further accelerated progress in analyzing, sequencing, and mapping sections of the genome. The advisability of granting private biotechnology firms the right to patent specific genes and DNA sequences was hotly debated. In April 1992 Watson resigned as director of the project to express his vehement disapproval of the NIH decision to patent human gene sequences. Later that year preliminary physical and genetic maps of the human genome were published.

Francis S. Collins of the NIH was named as the director of the HGP in April 1993, and international efforts to assist were underway in England, France, Germany, Japan, and other countries. In 1995, when Stanford University researchers released DNA chip technology that simultaneously analyzes genetic information representing thousands of genes, the development promised to speed the project to completion even before the anticipated date of 2005.

In 1995 investigators at the Institute for Genomic Research published the first complete genome sequence for any organism: the bacterium *Haemophilus influenzae*, with nearly 2 million genetic letters and 1,000 recognizable genes. In 1997 a yeast genome, *Saccharomyces cerevisiae*, composed of about 6,000 genes, was sequenced, and later that year the genome of the bacterium *Escherichia coli*, also known as *E. coli,* which contains approximately 4,600 genes, was sequenced.

In 1998 the genome of the first multicelled animal, the nematode worm *Caenorhabditis elegans*, was sequenced, containing approximately 18,000 genes. The following year the first complete sequence of a human chromosome was published by the HGP. In 2000 the genome of the fruit fly *Drosophila melanogaster*, which Morgan and his colleagues had used to study genetics nearly a century earlier, was sequenced by the private firm Celera Genomics. The fruit fly sequence contains about 13,000 genes, with many sequences matching already identified human genes that are associated with inherited disorders.

In 2000 the first draft of the human genome was announced, and it was published in 2001. Also in 2000 the first plant genome, *Arabidopsis thaliana*, was sequenced. This feat spurred research in plant biology and agriculture. Although tomatoes that had been genetically engineered for longer shelf life had been marketed during the mid-1990s, agricultural researchers began to see new possibilities to enhance crops and food products. For example, in 2000 plant geneticists developed genetically engineered rice that manufactured its own vitamin A. Many researchers believe the genetically enhanced strain of rice holds great promise in terms of preventing vitamin A deficiency in developing countries.

The 2001 publication of the human genome estimated that humans have between 30,000 and 35,000 genes. The HGP was completed in 2003. The same year the Cold Spring Harbor Laboratory held educational events to commemorate and celebrate the fiftieth anniversary of the discovery of the double helical structure of DNA. In October 2004 human gene count estimates were revised downward to between 20,000 and 25,000. During 2005 and 2006 sequencing of more than ten human chromosomes was completed, including the human X chromosome, which is one of the two sex chromosomes; the other is the Y chromosome. In October 2006 Roger D. Kornberg, an American biochemist at Stanford University, was awarded the Nobel Prize in Chemistry for determining the intricate way in which information in the DNA of a gene is copied to provide the instructions for building and running a living cell. The postgenomic era began with a firestorm of controversies about the direction of genetic research, human cloning, stem cell research, and genetically modified food and crops.

CHAPTER 2
UNDERSTANDING GENETICS

The study of genetics—the inheritance of traits in living organisms—is a basic concept in biology. The same processes that provide the mechanism for organisms to pass genetic information to their offspring lead to the gradual change of species over time, which in turn produces biodiversity (the variety of life and the genetic differences among living organisms) and the evolution of new species. An understanding of genetics is becoming increasingly important as genetics research and technology—and the controversies surrounding them—gain greater influence on social trends and individual lives.

Rapid and revolutionary advances in genetics, medicine, and biotechnology have created new opportunities as well as a complex and far-reaching range of legal, ethical, and social issues. Genetic testing to screen for disease susceptibility, moral questions about the cloning of human organs, and the debate over genetically modified foods affect many lives. To fully appreciate the benefits, risks, and ramifications of genetics research and to critically evaluate the related issues raised by ethicists, scientists, and other stakeholders, it is vital to understand the basics of genetics, a discipline that integrates biology, mathematics, sociology, medicine, and public health.

To understand genetics, the development of organisms, and the diversity of species, it is important to learn how deoxyribonucleic acid (DNA) functions as the information-bearing molecule of living organisms. This chapter contains definitions of basic genetics terms such as *DNA*, *chromosomes*, and *proteins*, as well as an introduction to genetic concepts and processes, including reproduction and inheritance.

DNA CODES FOR PROTEINS

From the perspective of genetics, the DNA molecule has two major attributes. The first is that it is able to replicate—that is, to make an exact copy of itself that can be passed to another cell, thereby conveying its precise genetic characteristics. Figure 2.1 is a diagram that shows how DNA replication produces two completely new and identical daughter strands of DNA. The second critical attribute is that it stores detailed instructions to manufacture specific proteins—molecules that are essential to every aspect of life. DNA is a blueprint or template for making proteins, and much of the behavior and physiology (life processes and functions) of a living organism depends on the repertoire of proteins its DNA molecules know how to manufacture.

The function of DNA depends on its structure. The double strand of DNA is composed of individual building blocks called nucleotides that are paired and connected by chemical bonds. A nucleotide contains one of four nitrogenous bases: the purines (nitrogenous bases with two rings) adenine (A) and guanine (G), or the pyrimidines (nitrogenous bases with one ring; pyrimidines are smaller than purines) cytosine (C) and thymine (T). The two strands of DNA lie side by side to create a predictable sequence of nitrogenous base pairs. (See Figure 2.2.) A stable DNA structure is formed when the two strands are a constant distance apart, which can occur only when a purine (A or G) on one strand is paired with a pyrimidine (T or C) on the other strand. A generally pairs with T, and G generally pairs with C.

Proteins are molecules that perform all the chemical reactions necessary for life and provide structure and shape to cells. The properties of each protein depend primarily on its shape, which is determined by the sequence of its building blocks, known as amino acids. Proteins may be tough like collagen, the most abundant protein in the human body, or they may be stretchy like elastin, a protein that mixes with collagen to make softer, more flexible tissues such as skin. Figure 2.3 shows the shapes of four types of protein structures.

Proteins can act as structural components, building the tissues of the body. For example, some of the proteins in an egg include a bond that acts like an axle, allowing

FIGURE 2.1

DNA Replication

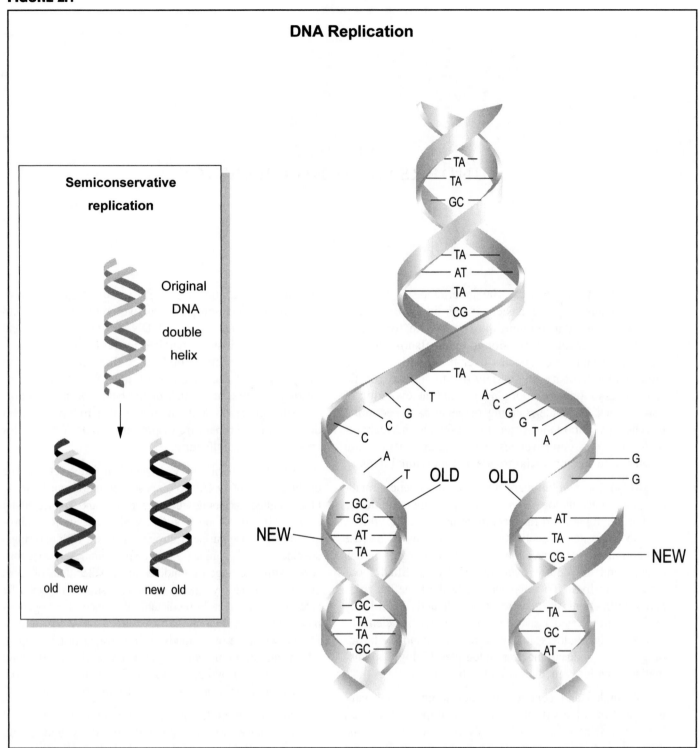

DNA replication. *Argosy Publishing, Thomson Gale.*

other parts of the molecule to spin around like wheels. However, when the egg is heated, these bonds break or denature, locking the "wheels" of the molecule in place. This is why an egg gets hard when you cook it.

Enzymes such as lactase, which helps in the digestion of lactose (milk sugar), and hormones such as insulin (a hormone that regulates carbohydrate metabolism by controlling blood sugar levels) are proteins that act to facilitate and direct chemical reactions. Defense proteins, which are able to combat invasion by bacteria or viruses, are embedded in the walls of cells and act as channels, determining which substances to let into the cell and which to block. Some bacteria know how to make proteins that protect them from antibiotics (substances such as penicillin and streptomycin that inhibit the growth of

FIGURE 2.2

Deoxyribonucleic acid (DNA)

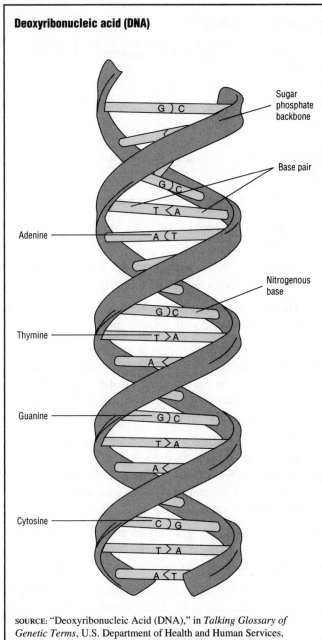

Sugar phosphate backbone

Base pair

Adenine

Nitrogenous base

Thymine

Guanine

Cytosine

SOURCE: "Deoxyribonucleic Acid (DNA)," in *Talking Glossary of Genetic Terms*, U.S. Department of Health and Human Services, National Institutes of Health, National Human Genome Research Institute, Division of Intramural Research, http://www.genome.gov/Pages/Hyperion//DIR/VIP/Glossary/Illustration/dna.shtml (accessed October 9, 2006)

FIGURE 2.3

Four protein structures

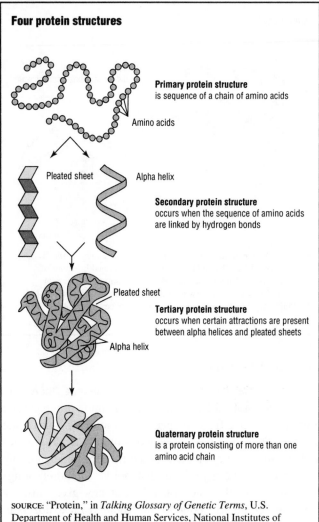

Primary protein structure
is sequence of a chain of amino acids

Amino acids

Pleated sheet

Alpha helix

Secondary protein structure
occurs when the sequence of amino acids are linked by hydrogen bonds

Pleated sheet

Tertiary protein structure
occurs when certain attractions are present between alpha helices and pleated sheets

Alpha helix

Quaternary protein structure
is a protein consisting of more than one amino acid chain

SOURCE: "Protein," in *Talking Glossary of Genetic Terms*, U.S. Department of Health and Human Services, National Institutes of Health, National Human Genome Research Institute, Division of Intramural Research, http://www.genome.gov/Pages/Hyperion//DIR/VIP/Glossary/Illustration/protein.shtml (accessed December 13, 2006)

or destroy microorganisms), whereas the human immune system can make proteins that target bacteria or other germs for destruction. Many essential biological processes depend on the highly specific functions of proteins.

Proteins and Amino Acids

All proteins are composed of building blocks called amino acids. (See Figure 2.4.) There are twenty different kinds of amino acids, and each has a slightly different chemical composition. The structure and function of each protein depends on its amino acid sequence—in a protein containing a hundred amino acids, a change in a single one may dramatically affect the function of the protein. Amino acids are small molecular groups that act like jigsaw puzzle pieces, linking together in a chain to make up the protein. Each amino acid links to its neighbor with a special kind of covalent bond (covalent bonds hold atoms together) called a peptide bond. Many amino acids link together, side by side, to make a protein. Figure 2.5 is a diagram of a protein structure. Proteins range in length from 50 to 500 amino acids, linked head to tail.

In addition to their ability to form peptide bonds to their neighbors, amino acids also contain molecular appendages called side groups. Depending on the particular atomic arrangement of the side group, neighboring amino acids experience different pushes and pulls as they attract or repel one another. The combination of the side-by-side peptide bond linking the amino acids into a chain, along with the extra influences of the side groups, twists a protein into a specific shape. This shape is called the protein's conformation, which determines how the protein interacts with other molecules.

FIGURE 2.4

Amino acids

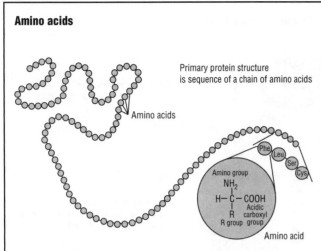

Primary protein structure is sequence of a chain of amino acids

Amino acids

Phe Leu Ser Cys

Amino group
NH₂
H – C – COOH
R   Acidic
R group   carboxyl
group   group

Amino acid

SOURCE: "Amino Acids," in *Talking Glossary of Genetic Terms*, U.S. Department of Health and Human Services, National Institutes of Health, National Human Genome Research Institute, Division of Intramural Research, http://www.genome.gov/Pages/Hyperion//DIR/VIP/Glossary/Illustration/amino_acid.shtml (accessed December 13, 2006)

FIGURE 2.5

A primary protein structure

Amino acids are linked head to tail, so that at one end there is a free amino group, and at the other a free carboxyl group. Proteins are typically 50–500 amino acids in length.

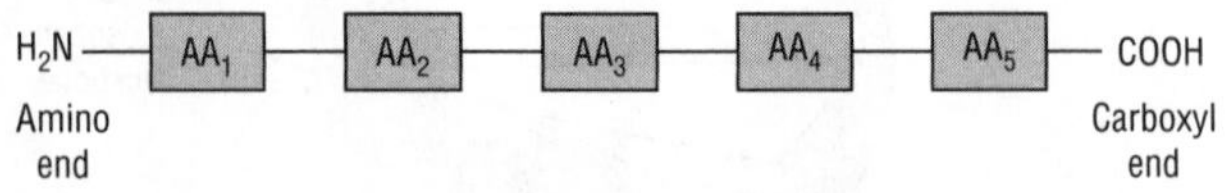

H₂N — AA₁ — AA₂ — AA₃ — AA₄ — AA₅ — COOH
Amino   Carboxyl
end   end

SOURCE: Adapted from Richard Robinson, ed., "Schematic Diagram of a Primary Protein Structure," in *Genetics*, Vol. 3, *K–P*, Macmillan Reference USA, 2002

1. A gene is triggered for expression—to synthesize a protein.

2. Half of the gene is copied into a single strand of mRNA in a process called transcription.

3. The mRNA anchors to a ribosome, an organelle (or membrane-bound cell compartment) where protein synthesis occurs.

4. Each sequence of three bases on the mRNA, called a codon, uses the right type of transfer RNA (tRNA) to pick up a corresponding amino acid from the cell. (See Figure 2.8.)

5. A string of amino acids is assembled on the ribosome, side by side, in the same order as the codons of the mRNA, which, in turn, correspond to the sequence of bases from the original DNA molecule in a process called translation.

6. When the codons have all been read and the entire sequence of amino acids has been assembled, the protein is released to twist into its final form.

First, the DNA molecule receives a trigger telling it to express a particular gene. Many influences may trigger gene expression, including chemical signals from hormones and energetic signals from light or other electromagnetic energy. For example, the spiral backbone of the DNA molecule can actually carry pulsed electrical signals that participate in activating gene expression.

Once a gene has been triggered for expression, a special enzyme system causes the DNA's double spiral to spring apart between the beginning and end of the gene sequence. The process is similar to a zipper with teeth that remain connected above and below a certain area, but pop open to create a gap along part of the zipper's length. At this point, the DNA base pairs that make up the gene sequence are separated. DNA replication is based on the understanding that any exposed nucleotide thymine (T) will pick up an adenine (A) and vice versa, while an exposed cytosine (C) will connect to an available guanine (G) and vice versa. Here the same process takes place except that instead of unzipping and copying the entire

Genes and Proteins

Along the length of a DNA molecule there are regions that hold the instructions to manufacture specific proteins—a specific sequence of amino acids linked side by side. These regions are called protein-encoding genes and are an essential element of the modern understanding of genetics. Like a jukebox that holds a hundred songs but only plays the one you select at any given time, a DNA molecule can contain anywhere from a dozen to several thousand of these protein-encoding genes. However, as with the jukebox, at any given time only some of these genes will be expressed—that is, switched on to actively produce the protein they know how to make.

The protein-synthesizing instructions present in DNA are interpreted and acted on by ribonucleic acid (RNA). RNA, as its name suggests, is similar to DNA, except that the sugar in RNA is ribose (instead of deoxyribose), the base uracil (U) replaces thymine (T), and RNA molecules are usually single stranded and shorter than DNA molecules. (See Figure 2.6.) RNA is used to transcribe and translate the genetic code contained in DNA. Transcription is the process by which a molecule of messenger RNA (mRNA) is made, and translation is the synthesis of a protein using mRNA code. Figure 2.7 shows transcription of mRNA and how mRNA is involved in protein synthesis (translation).

Genetic Synthesis of Proteins

How does a gene make a protein? The process of protein synthesis is quite complex. This overview describes the basic sequence of protein synthesis, which includes the following steps:

FIGURE 2.6

Comparison of ribonucleic acid (RNA) with deoxyribonucleic acid (DNA)

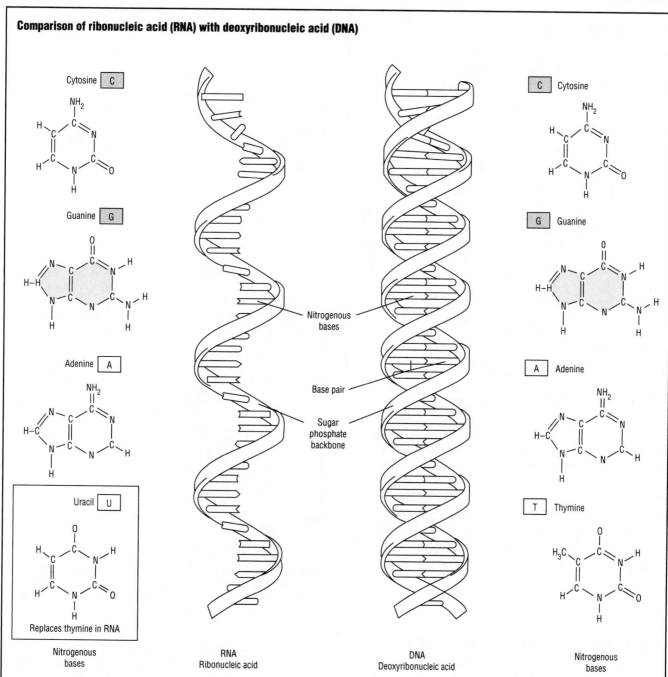

SOURCE: Adapted from "Ribonucleic Acid (RNA)," in *Talking Glossary of Genetic Terms*, U.S. Department of Health and Human Services, National Institutes of Health, National Human Genome Research Institute, Division of Intramural Research, http://www.genome.gov/Pages/Hyperion//DIR/VIP/Glossary/Illustration/rna.shtml (accessed December 13, 2006).

length of the DNA molecule, only the region between the beginning and end of the gene is copied. Instead of making a new double spiral, only one side of the gene is replicated, creating a special, single strand of bases called mRNA.

Once the mRNA has made a copy of one side of the gene, it separates from the DNA molecule. The DNA returns to its original state and the mRNA molecule breaks away, eventually connecting to an organelle in the cell called a ribosome. (Figure 2.9 is a drawing of a ribosome.) Here it anchors to another type of RNA called ribosomal

RNA (rRNA). This is where the actual protein synthesis takes place. Using a special genetic coding system, each sequence of three bases on the mRNA copied from the gene is used to catch a corresponding amino acid floating inside the cell. These sequences of three bases are the codons, and they link to tRNA. A different form of tRNA is used to catch each different type of amino acid. The tRNA then catches the appropriate amino acid.

One after another, the mRNA's codons cause the corresponding tRNAs to be captured and linked, side by side,

FIGURE 2.7

Messenger RNA (mRNA)

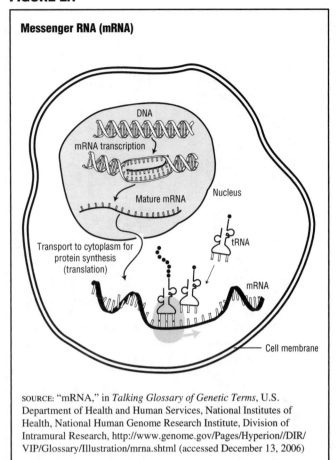

SOURCE: "mRNA," in *Talking Glossary of Genetic Terms*, U.S. Department of Health and Human Services, National Institutes of Health, National Human Genome Research Institute, Division of Intramural Research, http://www.genome.gov/Pages/Hyperion//DIR/VIP/Glossary/Illustration/mrna.shtml (accessed December 13, 2006)

using the peptide bonds to connect them. When the last codon has been read and the entire peptide sequence is complete, the newly formed protein molecule is released from the ribosome. When this happens, all the side groups are able to interact, twisting the protein into its final shape. In this way, a protein-encoding gene is able to manufacture a protein from a series of DNA bases.

THE CELL IS THE BASIC UNIT OF LIFE

Ever since Matthias Jakob Schleiden (in 1838) and Theodor Schwann (in 1839) put forth their theories that all plants and animals are composed of cells, there has been continuous refinement of cell theory. The early view that cells were made up of protoplasm (a jellylike substance) has given way to the more sophisticated understanding that cells are highly complex organizations of even smaller molecules and substructures. Cytology—the study of the formation, structure, and function of cells—has benefited from ever-improving technology, including powerful microscopes that enable researchers to identify the organelles (component parts of the cell) and determine their roles in inheritance.

Cells are the basic units and building blocks of nearly every organism. (One exception is viruses, which are simple organisms that are not composed of cells.) Each

FIGURE 2.8

Codon

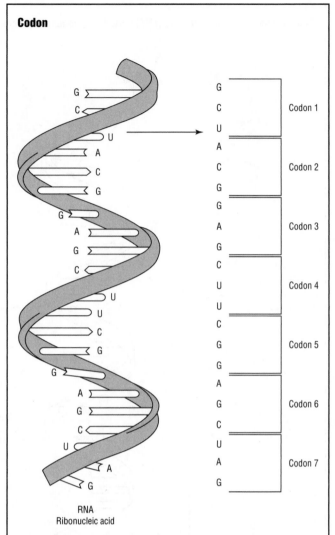

RNA
Ribonucleic acid

SOURCE: Adapted from "Codon," in *Talking Glossary of Genetic Terms*, U.S. Department of Health and Human Services, National Institutes of Health, National Human Genome Research Institute, Division of Intramural Research, http://www.genome.gov/Pages/Hyperion//DIR/VIP/Glossary/Illustration/codon.shtml (accessed December 13, 2006)

cell of an organism contains the same genetic information, which is passed on faithfully when cells divide. Different types of cells arise because they use different parts of the information, as determined by the cell's history and the immediate environment. Different cell types may be organized into tissues and organs.

Cell Structure and Function

Plants and animals, as well as other organisms such as fungi, are composed of eukaryotic cells, or eukaryotes, because they have nuclei and membrane-bound structures known as organelles. In eukaryotic cells the organelles within the cell sustain, support, and protect it, creating a barrier between the cell and its environment, acting to build and repair cell parts, storing and releasing energy, transporting material, disposing of waste, and increasing in number.

FIGURE 2.9

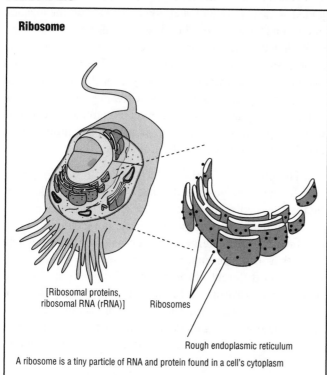

Ribosome

[Ribosomal proteins, ribosomal RNA (rRNA)]

Ribosomes

Rough endoplasmic reticulum

A ribosome is a tiny particle of RNA and protein found in a cell's cytoplasm

SOURCE: Adapted from "Ribosome," in *Talking Glossary of Genetic Terms*, U.S. Department of Health and Human Services, National Institutes of Health, National Human Genome Research Institute, Division of Intramural Research, http://www.genome.gov/Pages/Hyperion//DIR/VIP/Glossary/Illustration/ribosomes.shtml (accessed December 13, 2006)

FIGURE 2.10

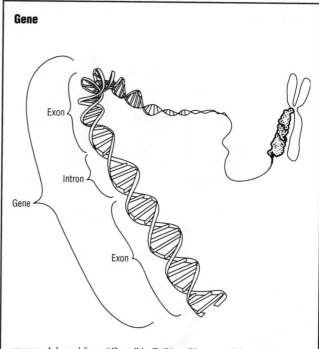

Gene

Exon

Intron

Gene

Exon

SOURCE: Adapted from "Gene," in *Talking Glossary of Genetic Terms*, U.S. Department of Health and Human Services, National Institutes of Health, National Human Genome Research Institute, Division of Intramural Research, http://www.genome.gov/Pages/Hyperion//DIR/VIP/Glossary/Illustration/gene.shtml (accessed October 16, 2006)

Each organelle functions like an organ system for the cell. For example, the nucleus is the command center, masterminding protein synthesis within the cell. The ribosomes work as protein factories, the Golgi apparatus is a protein sorter, and the endoplasmic reticulum operates as a protein processor. Lysosomes and peroxisomes serve as the cell's digestive system, and mitochondria convert energy in the cell. (See Figure 1.1 in chapter 1.) The surface membrane of the cell acts like skin, selectively permitting molecules in and out of the cell.

The nuclei of eukaryotes contain the chromosomes, chains of genetic material coded in DNA. The threadlike chromosomes are contained in the nucleus of a typical animal cell. Genes are segments of DNA that carry a basic unit of hereditary information in coded form. (See Figure 2.10.) They contain instructions for making proteins.

The eukaryotic chromosome is composed of chromatin (a combination of nuclear DNA and protein) and contains a linear array of genes. It is visible just before and during cell division. (See Figure 2.11.) Human cells normally contain twenty-three pairs of chromosomes, or a total of forty-six chromosomes, that may be examined using a process known as karyotyping, the organization of a standard picture of the chromosomes. Figure 2.12 is a karyotype—a photo of an individual's chromosomes.

Cells without nuclei, such as bacteria and blue-green algae, are called prokaryotic cells or prokaryotes. Prokaryotic cells are smaller than eukaryotic cells, contain less genetic information, and are able to grow and divide more quickly. They perform these functions without organelles. A prokaryotic cell's DNA is not contained in one location; instead, it floats in different regions of the cell. The DNA of prokaryotes also contains jumping genes that are able to bind to other genes and transfer gene sequences from one site of a chromosome to another.

GROWTH AND REPRODUCTION

The growth of an organism occurs as a result of cell division in a process known as mitosis. Many cells are relatively short lived, and mitosis allows for regular renewal of these cells. It is also the process that generates the millions of cells needed to grow an organism, or in the case of a human being, the trillions of cells needed to grow from birth to adulthood.

Mitosis is a continuous process that occurs in several stages. Between cell divisions, the cells are in interphase, during which there is cell growth, and the genetic material—DNA—contained in the chromosomes is duplicated so that when the cell divides, each new cell has a full-scale version of the same genetic material. The process of mitosis involves exact duplication—gene by gene—of the cell's chromosomal material and a systematic method for evenly distributing

FIGURE 2.11

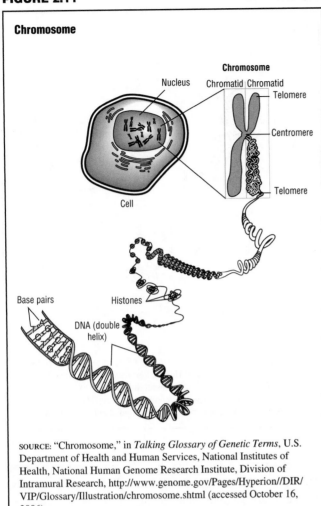

Chromosome

SOURCE: "Chromosome," in *Talking Glossary of Genetic Terms*, U.S. Department of Health and Human Services, National Institutes of Health, National Human Genome Research Institute, Division of Intramural Research, http://www.genome.gov/Pages/Hyperion//DIR/VIP/Glossary/Illustration/chromosome.shtml (accessed October 16, 2006)

spot called the centromere. Figure 2.11 shows chromosomes joined at the centromere, and Figure 2.13 shows the location of the centromere in the chromosome. Pairing up along their entire lengths, they are able to exchange genetic material in a process known as crossing over. Figure 2.14 shows homologous chromosomes crossing over during meiosis to create new gene combinations. Crossing over results in much of the genetic variation observed among parents and their offspring. The pairing of homologous chromosomes and crossing over occur only in meiosis.

The process of meiosis also creates another opportunity to generate genetic diversity. During one phase of meiosis, called metaphase, the arrangement of each pair of homologous chromosomes is random, and different combinations of maternal and paternal chromosomes line up with varying orientations to create new gene combinations on different chromosomes. This action is called independent assortment. Figure 2.15 shows the process of meiosis in an organism with six chromosomes. Because recombination and independent assortment of parental chromosomes takes place during meiosis, the daughter cells are not genetically identical to one another.

Gamete Formation

In male animals gamete formation, known as spermatogenesis, begins at puberty and takes place in the testes. Spermatogenesis involves a sequence of events that begin with the mitosis of primary germ cells to produce primary spermatocytes. (See Figure 2.16.) Four primary spermatocytes undergo two meiotic divisions and as they undergo spermatogenesis they lose much of their cytoplasm and develop the spermatozoa's characteristic tail, the motor apparatus that provides the propulsion necessary to reach the egg cell. Like oocytes (egg cells) produced by the female, the mature sperm cells will be haploid, possessing just one copy of each chromosome.

In female animals gamete formation, known as oogenesis, takes place in the ovaries. Primary oogocytes are produced by mitosis in the fetus before birth. Unlike the male, which continues to produce sperm cells throughout life, in the female the total number of eggs ever to be produced is present at birth. At birth, or shortly before, meiosis begins and primary oocytes remain in the prophase of meiosis until puberty. At puberty the first meiotic division is completed, a diploid cell becomes two haploid daughter cells; one large cell becomes the secondary oocyte and the other the first polar body. The secondary oocyte undergoes meiosis a second time but the meiosis does not continue to completion without fertilization. Figure 2.17 shows the sequence of events leading to the production of a mature ovum.

Fertilization

Like gamete formation, fertilization is a process, as opposed to a single event. It begins when the sperm and

this material. It concludes with the physical division known as cytokinesis, when the identical chromosomes pull apart and each heads for the nucleus of one of the new daughter cells. Mitosis occurs in all eukaryotic cells, except the gametes (sperm and egg), and always produces genetically identical daughter cells with a complete set of chromosomes.

If, however, mitosis occurred in the gametes, then when fertilization—the joining of sperm and egg—took place, the offspring would receive a double dose of hereditary information. To prevent this from occurring, the gametes undergo a process of reduction division known as meiosis. Meiosis reduces the number of chromosomes in the gametes by half, so that when fertilization occurs the normal number of chromosomes is restored. For example, in humans the gametes produced by meiosis are haploid—they have just one copy of each of the twenty-three chromosomes. Besides preventing the number of chromosomes from doubling with each successive generation, meiosis also provides genetic diversity in offspring.

During meiosis the chromosomal material replicates and concentrates itself into homologous chromosomes (doubled chromosomes), each of which is joined at a central

FIGURE 2.12

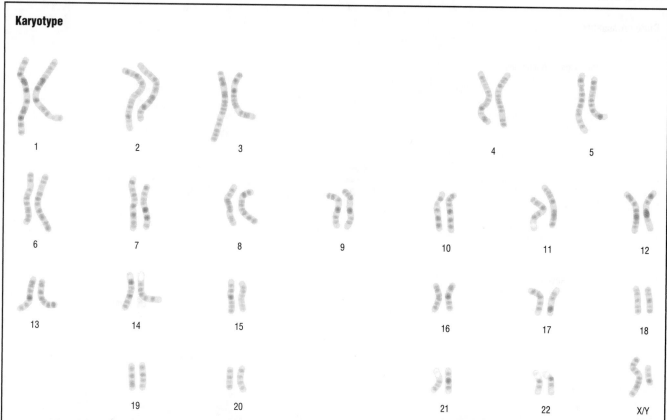

Karyotype

1 2 3 4 5

6 7 8 9 10 11 12

13 14 15 16 17 18

19 20 21 22 X/Y

SOURCE: Adapted from "Karyotype," in *Talking Glossary of Genetic Terms*, U.S. Department of Health and Human Services, National Institutes of Health, National Human Genome Research Institute, Division of Intramural Research, http://www.genome.gov/Pages/Hyperion//DIR/VIP/Glossary/Illustration/karyotype.shtml (accessed October 16, 2006)

FIGURE 2.13

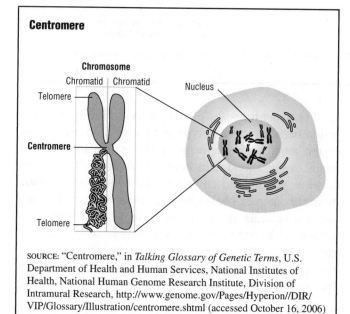

Centromere

Chromosome

Chromatid | Chromatid

Telomere

Nucleus

Centromere

Telomere

SOURCE: "Centromere," in *Talking Glossary of Genetic Terms*, U.S. Department of Health and Human Services, National Institutes of Health, National Human Genome Research Institute, Division of Intramural Research, http://www.genome.gov/Pages/Hyperion//DIR/VIP/Glossary/Illustration/centromere.shtml (accessed October 16, 2006)

Of the millions of sperm released during an ejaculation, less than 1% survives to reach the egg. Of the few hundred sperm that reach the egg, only one will successfully fertilize it. While the sperm are in the female reproductive tract, swimming toward the egg, they undergo a process known as capacitation, during which they acquire the capacity to fertilize the egg. As the sperm approach the egg they become hyperactivated, and in a frenzy of mechanical energy the sperm attempt to burrow their way through the outer shell of the egg called the zona pellucida.

The cap of the sperm, known as the acrosome, contains enzymes that are crucial for fertilization. These acrosomal enzymes dissolve the zona pellucida by making a tiny hole in it, so that one sperm can swim through and reach the surface of the egg. At this time, the egg transforms the zona pellucida by creating an impenetrable barrier, so that no other sperm may enter.

Sperm penetration triggers the second meiotic division of the egg. With this division the chromosomes of the sperm and egg form a single nucleus. The resulting cell—the first cell of an entirely new organism—is called a zygote. The zygote then divides into two cells, which, in turn, continue to divide rapidly, producing a ball of cells now called the blastocyst. The blastocyst is an early

egg first come into contact and fuse together and culminates with the intermingling of two sets of haploid genes to reconstitute a diploid cell with the potential to become a new organism.

FIGURE 2.14

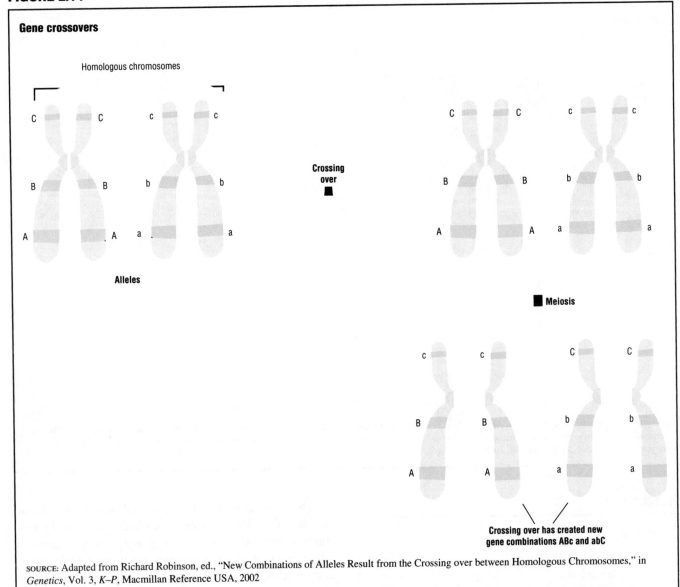

Gene crossovers

SOURCE: Adapted from Richard Robinson, ed., "New Combinations of Alleles Result from the Crossing over between Homologous Chromosomes," in *Genetics*, Vol. 3, *K–P*, Macmillan Reference USA, 2002

stage of embryogenesis, the process that describes the development of the fertilized egg as it becomes an embryo. In humans the developing baby is considered an embryo until the end of the eighth week of pregnancy.

GENETIC INHERITANCE

For inheritance of simple genetic traits, the two inherited copies of a gene determine the phenotype (the observable characteristic) for that trait. When genes for a particular trait exist in two or more different forms that may differ among individuals and populations, they are called alleles. For example, brown and blue eye colors are different alleles for eye color. For every gene, the offspring receives two alleles, one from each parent. The combination of inherited alleles is the genotype of the organism, and its expression—the observable characteristic—is its phenotype. Figure 2.18 is a graphic example of phenotype.

For many traits the phenotype is a result of an interaction between the genotype and the environment. Some of the most readily apparent traits in humans, such as height, weight, and skin color, result from interactions between genetic and environmental factors. In addition, there are complex phenotypes that involve multiple gene-encoded proteins; the alleles of these particular genes are influenced by other factors, either genetic or environmental. So while the presence of certain genes indicates susceptibility or likelihood to develop a certain trait, it does not guarantee expression of the trait.

For a specific trait some alleles may be dominant while others are recessive. The phenotype of a dominant allele is always expressed, while the phenotype of a recessive allele is expressed only when both alleles are recessive. Recessive genes continue to pass from generation to generation, but they are only expressed in individuals who do not inherit a copy of the dominant gene for the specific trait.

FIGURE 2.15

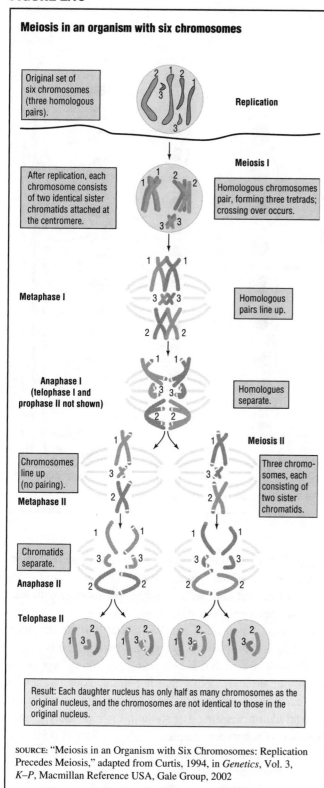

Meiosis in an organism with six chromosomes

Original set of six chromosomes (three homologous pairs).

Replication

After replication, each chromosome consists of two identical sister chromatids attached at the centromere.

Meiosis I

Homologous chromosomes pair, forming three tretrads; crossing over occurs.

Metaphase I

Homologous pairs line up.

Anaphase I (telophase I and prophase II not shown)

Homologues separate.

Meiosis II

Chromosomes line up (no pairing).

Metaphase II

Three chromosomes, each consisting of two sister chromatids.

Chromatids separate.

Anaphase II

Telophase II

Result: Each daughter nucleus has only half as many chromosomes as the original nucleus, and the chromosomes are not identical to those in the original nucleus.

SOURCE: "Meiosis in an Organism with Six Chromosomes: Replication Precedes Meiosis," adapted from Curtis, 1994, in *Genetics*, Vol. 3, *K–P*, Macmillan Reference USA, Gale Group, 2002

Figure 2.19 shows the inheritance of a recessive trait, in this example a recessive mutation in mice that produces an albino offspring from black parents.

There are also some instances, known as incomplete dominance, when one allele is not completely dominant over the other and the resulting phenotype is a blend of both traits. Skin color in humans is an example of a trait often governed by incomplete dominance, with offspring appearing to be a blend of the skin tones of each parent. Furthermore, some traits are determined by a combination of several genes (multigenic or polygenic), and the resulting phenotype is determined by the final combination of alleles of all the genes that govern the particular trait.

Some multigenic traits are governed by many genes, each contributing equally to the expression of the trait. In such instances a defect in a single gene pair may not have a significant impact on expression of the trait. Other multigenic traits are predominantly directed by one major gene pair and only mildly influenced by the effects of other gene pairs. For these traits the impact of a defective gene pair depends on whether it is the major pair governing expression of the trait or one of the minor pairs influencing its expression.

A range of other factors enters into whether a trait will be evidenced and the extent to which it is expressed. For example, different individuals may express a trait with different levels of severity. This phenomenon is known as variable expressivity.

Determining Genetic Probabilities of Inheritance

Conventionally, geneticists use uppercase letters to represent dominant alleles and lowercase letters to stand for recessive alleles. An organism with a pair of identical alleles for a trait is described as a homozygote, or homozygous for that particular trait. When organisms are homozygous for a dominant trait, all uppercase letters symbolize the trait, while those that are homozygous for a recessive trait are represented by all lowercase letters. A heterozygote is an organism with different alleles for a trait, one donated from each parent, and when one is dominant and the other is recessive the trait is shown using a combination of uppercase and lowercase letters.

Even though the combination of alleles is a random event, it is possible to predict the probability that an offspring will have the same or a different phenotype from its parents when the genotypes with respect to the specific trait of both parents and the phenotype associated with each possible combination of alleles are known. The formula used to determine genetic possibilities was developed by the English geneticist Reginald Crundall Punnett. The Punnett square is a grid configuration that depicts genotype and phenotype. In Punnett squares the genotypes of parents are represented with four letters. There are two alleles for each trait. Genotypes of haploid gametes are represented with two letters. Gametes will contain one allele for each trait in every possible combination, and all possible fertilizations are calculated. Punnett squares are used to compute the

FIGURE 2.16

Process of spermatogenesis

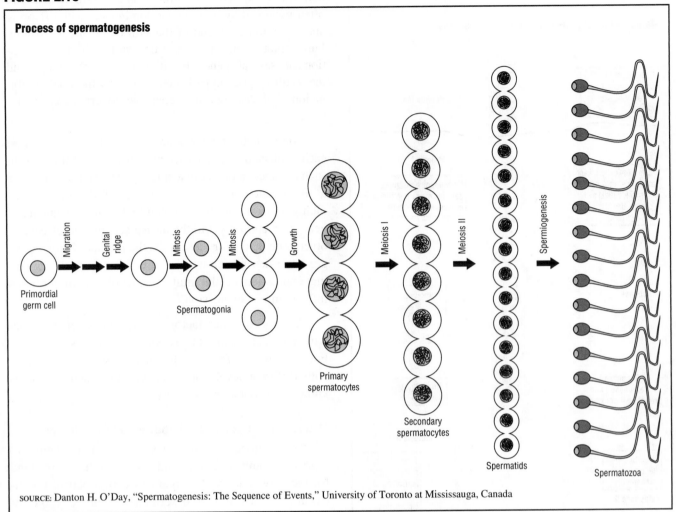

SOURCE: Danton H. O'Day, "Spermatogenesis: The Sequence of Events," University of Toronto at Mississauga, Canada

FIGURE 2.17

Process of oogenesis

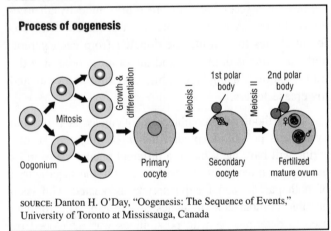

SOURCE: Danton H. O'Day, "Oogenesis: The Sequence of Events," University of Toronto at Mississauga, Canada

cross of a single gene or two genes and their alleles, but they become extremely complicated when used to predict the offspring of more than two alleles.

To construct a Punnett square, the alleles in the gametes of one parent are placed along the top, and the alleles in the gametes of the other parent are placed along the left side of the grid. The number of characteristics considered determines the size of the Punnett square. A monohybrid cross looks at one trait and has a two-by-two structure, with four possible genotypes resulting from the cross. A dihybrid cross looks at two traits and has a four-by-four structure, providing sixteen possible genotypes. Figure 2.20 is a simple Punnett square that shows one parent homozygous for the AA allele and another that is heterozygous (Aa), with the four possible offspring: 50% AA and 50% Aa. This is an example of offspring that show just one phenotype even though two genotypes are present. Figure 2.21 is a Punnett square that shows the predicted patterns of recessive inheritance. When both parents are carriers of the trait and dominant inheritance, the offspring is likely to arise from an affected parent and a normal parent.

Mitochondrial Inheritance

Mitochondria are the organelles involved in cellular metabolism and energy production and conversion. Mitochondria are inherited exclusively from the mother and contain their own DNA, known as mtDNA, some of which multiplies during the organism's growth and development

FIGURE 2.18

Phenotype

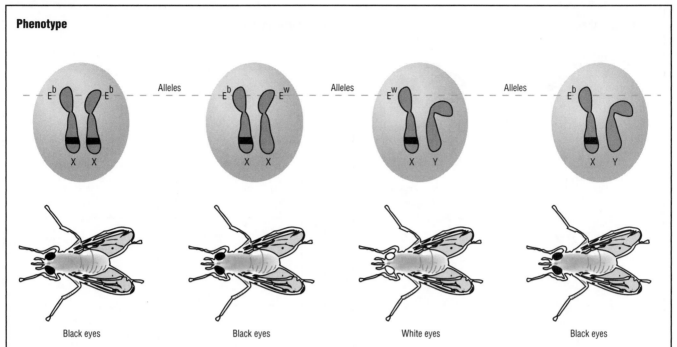

SOURCE: "Phenotype," in *Talking Glossary of Genetic Terms*, U.S. Department of Health and Human Services, National Institutes of Health, National Human Genome Research Institute, Division of Intramural Research, http://www.genome.gov/Pages/Hyperion//DIR/VIP/Glossary/Illustration/phenotype .shtml (accessed October 16, 2006)

and as such are more susceptible to mutation (changes in DNA sequence). They also acquire it more quickly than other DNA. The fact that mitochondria contain their own DNA has prompted scientists to speculate that they originally existed as independent one-celled organisms that over time developed interdependent relationships with more complex, eukaryotic cells.

Mitochondrial inheritance looks much like Mendelian inheritance (or genetic inheritance as described by Gregor Mendel's laws), with two important exceptions. First, all maternal offspring are usually affected, whereas even in autosomal dominant disorders (those not related to the sex genes) only 50% of offspring are expected to be affected. (See Figure 2.22.) Second, mitochondrially inherited traits are never passed through the male parent. Males are as likely to be affected as females, but their offspring are not at risk. In other words, when there is a mutation in a mitochondrial gene, it is passed from a mother to all of her children; sons will not pass it on, but daughters will pass it on to all their children.

Mutations in mtDNA have been linked to the development of disease in humans. Leber's hereditary optic neuropathy, a painless loss of vision that afflicts people between the ages of twelve and thirty, was the first human disease to be associated with a mutation in mtDNA. Many diseases linked to mtDNA affect the nervous system, heart or skeletal muscles, liver, or kidneys—sites of energy usage.

DETERMINING GENDER

From the moment of fertilization, the new organism has been assigned a gender, and its growth will proceed to develop either a male or a female organism. The first clues that prompted scientists to consider that the determination of gender was influenced by genetics came from two key observations. The first was the fact that there is a general tendency toward a one-to-one ratio of males to females in all species. The second was the realization that the determination of gender, or sex, followed the principles of Mendelian genetics—gender was predictable as expected when individuals pure for a recessive trait were crossed with individuals that were hybrid.

The determination of gender occurs in all complex organisms, but the processes vary, even among animals. In humans twenty-two of the twenty-three pairs of chromosomes are as likely to be found in males as in females. These twenty-two chromosomes are known as the autosomes, and the twenty-third pair is the sex chromosome. The sex chromosomes of females are identical and are called X chromosomes. In males the pair consists of an X chromosome and a smaller Y chromosome. (See Figure 2.23.)

Chromosome Theory of Sex Determination

The genetic influence on gender is called the chromosome theory of sex determination, which states that:

- Gender is determined by the sex chromosome.

- In females the sex chromosomes are identical—both are X chromosomes.

FIGURE 2.19

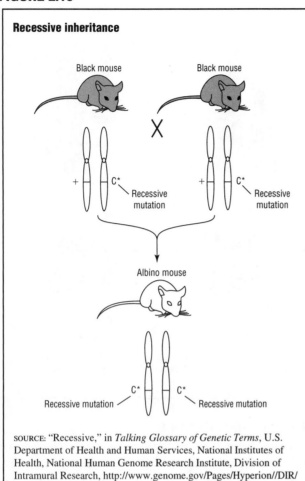

Recessive inheritance

SOURCE: "Recessive," in *Talking Glossary of Genetic Terms*, U.S. Department of Health and Human Services, National Institutes of Health, National Human Genome Research Institute, Division of Intramural Research, http://www.genome.gov/Pages/Hyperion//DIR/VIP/Glossary/Illustration/recessive.shtml (accessed October 16, 2006)

FIGURE 2.20

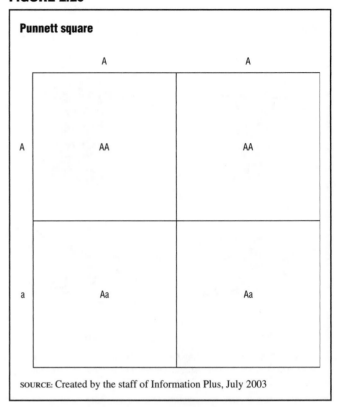

Punnett square

SOURCE: Created by the staff of Information Plus, July 2003

- Because females have an XX genotype, all egg cells contain an X chromosome.

- In males the sex chromosomes are not identical; one is X and one is Y.

- Because males have an XY genotype, half of all sperm cells contain an X chromosome and half contain a Y chromosome.

- Upon fertilization, the egg may receive either an X or a Y chromosome from the sperm. Because all egg cells contain an X chromosome, the determination of gender is wholly dependent on the chromosomal composition of the sperm. Sperm carrying the Y chromosome are known as androsperm; those containing the X chromosome are called gynosperm. If the sperm carries the Y chromosome, the offspring will be male (XY); if it carries an X chromosome, the offspring will be female (XX).

The determination of gender occurs at conception with the designation of chromosomal composition that is either XX or XY. However, a number of other genetic and environmental influences determine sex differentiation—the way in which the genetically predetermined gender becomes a reality. Differentiation translates the genetically coded message for gender into the physical traits, such as the hormones that influence development of male and female genitalia, body functions, and behaviors associated with gender identity.

Interestingly, humans have an inherent tendency toward female development. Research conducted during the 1940s and 1950s confirmed that in many animals individuals with just a single X chromosome developed as females, although in many instances they did not develop completely and were sterile (unable to reproduce). The absence of the Y chromosome results in female development, while the presence of the Y chromosome sets in motion the series of events that result in male development. These findings led to the premise that female development is the default option in the process of gender determination.

Distribution of Males and Females in the Population

Because human males produce equal numbers of sperm bearing either the X or the Y chromosome, and fertilization is a random event, then it stands to reason that in each generation equal numbers of males and females should be born. An examination of birth statistics in the United States and in other countries where reliable statistics have been compiled over time show that every year there are more births of males than females. For example, according to Joyce A. Martin et al., in "Births: Final Data for 2004" (*National Vital Statistics Reports*, September

FIGURE 2.21

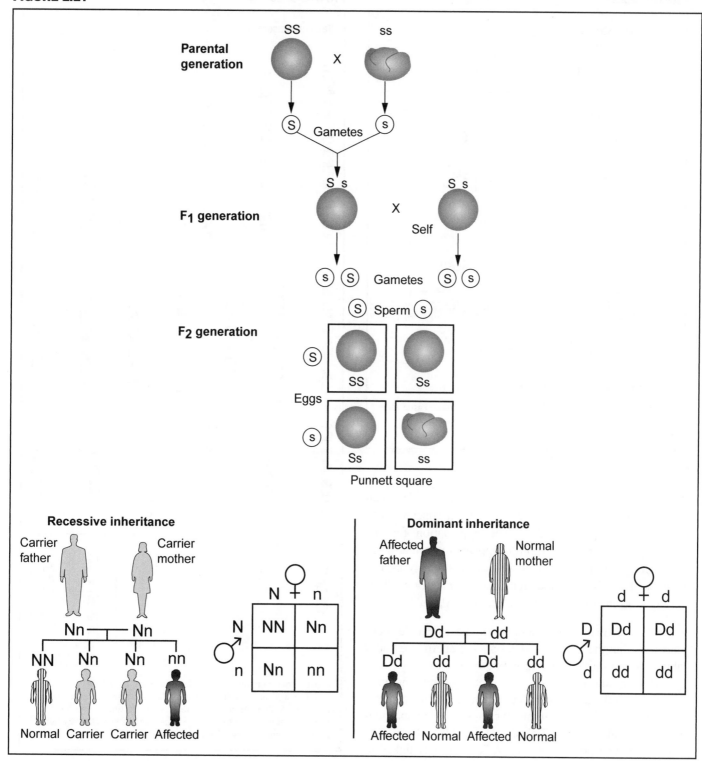

Punnett square with predicted patterns of recessive inheritance. *Argosy Publishing, Thomson Gale.*

29, 2006), there were 2,104,661 live male births, compared to 2,007,391 live female births, a ratio of 1,048 males per 1,000 females for births to mothers of all races in the United States in 2004. Table 2.1 shows that births by sex varied by race, from a low of 1,030 males per 1,000 females among Native Americans and Alaskan Natives to a high of 1,058 males per 1,000 females among Asians and Pacific Islanders. Martin and her collaborators report that the annual distribu-

tion of births by sex has remained essentially unchanged over the past sixty years, varying by less than 1%.

For years this difference was attributed to the idea that males were inherently stronger than females and better able to survive pregnancy and birth. This theory was dispelled when researchers found that nearly three times as many male fetuses spontaneously abort (dying

FIGURE 2.22

Inheritance patterns

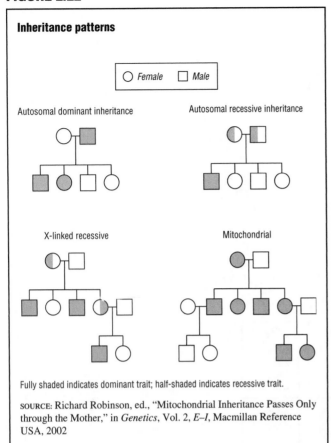

Fully shaded indicates dominant trait; half-shaded indicates recessive trait.

SOURCE: Richard Robinson, ed., "Mitochondrial Inheritance Passes Only through the Mother," in *Genetics*, Vol. 2, *E–I*, Macmillan Reference USA, 2002

FIGURE 2.23

Sex chromosomes

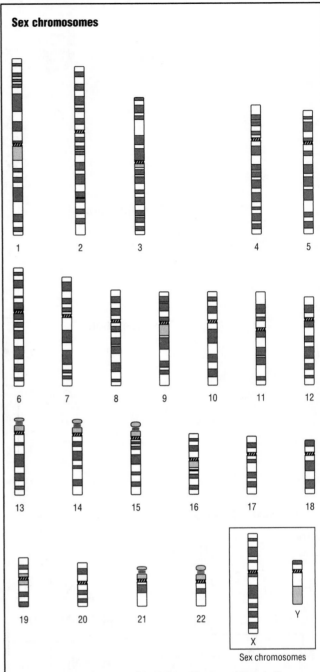

Sex chromosomes

SOURCE: "Sex Chromosomes," in *Talking Glossary of Genetic Terms*, U.S. Department of Health and Human Services, National Institutes of Health, National Human Genome Research Institute, Division of Intramural Research, http://www.genome.gov/glossary.cfm?key=sex%20chromosome (accessed December 8, 2006)

before birth). In fact, male life expectancy is less than female life expectancy at every age, from conception to adulthood. (See Table 2.2.) The explanation for the higher proportion of male births appears to be that more male offspring are conceived—possibly as many as 125 to every 100—because the prenatal death rate for males is so high; at birth the gap closes to about 105 to 100.

One explanation for the higher number of males conceived is that the smaller and stronger Y sperm are better able to swim quickly and successfully reach the egg cell. Along with androsperm size and mobility, environmental conditions influence gender determination and the chances of fetal survival. For example, the mother's age and general health are strongly linked to favorable outcomes of conception and pregnancy, and are less strongly linked to but are associated with gender. Younger mothers are more likely to conceive male offspring, by a ratio as high as 120 to 100, and unfavorable prenatal conditions such as poor health or maternal illness are more likely to compromise the survival of the male fetus than the female.

Sex-Linked Characteristics

The two sex chromosomes also differ in terms of the genes they contain, which relate to many traits other than gender. The Y chromosome is quite small and carries few

genes other than the one that determines male gender. One of the few confirmed traits linked to the Y chromosome is the hairy ear trait, a characteristic that is distinctive but unrelated to health. Because this trait is located exclusively on the Y chromosome, it only appears in males.

The X chromosome is larger and holds many genes that are as necessary for males as they are for females. The genes on the X chromosome are termed X-linked, and characteristics or conditions arising from these genes are called

TABLE 2.1

Total births, by race of mother and selected demographic characteristics, 2004

[Birth rates are live births per 1,000 population. Fertility rates are computed by relating total births, regardless of age of mother, to women aged 15–44 years. Total fertility rates are sums of birth rates for 5-year age groups multiplied by 5. Population estimated as of July 1. Mean age at first birth is the arithmetic average of the age of mothers at the time of the birth, computed directly from the frequency of first births by age of mother.]

Characteristic	All races	White	Black	American Indian or Alaska Native[a]	Asian or Pacific Islander
			Number		
Births	4,112,052	3,222,928	616,074	43,927	229,123
			Rate		
Birth rate	14.0	13.5	16.0	14.0	16.8
Fertility rate	66.3	66.1	67.6	58.9	67.1
Total fertility rate	2,045.5	2,054.5	2,032.5	1,734.5	1,897.5
Sex ratio[b]	1,048	1,050	1,039	1,030	1,058
All births			**Percent**		
Births to mothers under 20 years	10.3	9.3	17.1	17.9	3.4
4th-and higher-order births	11.0	10.4	15.1	19.5	6.4
Births to unmarried mothers	35.8	30.5	68.8	62.3	15.5
Mothers born in the 50 states and DC	75.8	78.0	84.4	94.7	17.3
			Mean		
Age of mother at first birth	25.2	25.4	22.7	21.8	28.4

[a]Includes births to Aleuts and Eskimos.
[b]Male live births per 1,000 female live births.
Notes: Race and Hispanic origin are reported separately on birth certificates. Race categories are consistent with the 1977 Office of Management and Budget (OMB) standards. Fifteen states reported multiple-race data for 2004. The multiple-race data for these states were bridged to the single-race categories of the 1977 OMB standards for comparability with other states. In this table all women (including Hispanic women) are classified only according to their race.

SOURCE: J. A. Martin, B. E. Hamilton, P. D. Sutton, S. J. Ventura, F. Menacker, and S. Kirmeyer, "Table 14. Total Number of Births, Rates (Birth, Fertility, and Total Fertility), and Percentage of Births with Selected Demographic Characteristics, by Race of Mother: United States, 2004," in "Births: Final Data for 2004," in *National Vital Statistics Reports*, vol. 55, no. 1, September 29, 2006, http://www.cdc.gov/nchs/data/nvsr/nvsr55/nvsr55_01.pdf (accessed October 16, 2006)

X-linked traits or conditions. Most people, male and female, likely have several so-called defective genes with the potential to produce harmful characteristics or conditions, but these genes are usually recessive and are not expressed in the phenotype unless they are combined with a similar recessive gene on the corresponding chromosome. For this to occur, both parents would have to contribute the same defective gene. Species are also protected from the harmful effects of single defective genes by virtue of the fact that most traits are multigenic (controlled by more than one gene).

Although X-linked dominant alleles affect males and females, males are more strongly affected because they inherit just one X chromosome and do not have a counterbalancing normal allele. Huntington's disease, an inherited neuropsychiatric disease that affects the body and mind, is an example of a disorder caused by an X-linked dominant allele. Males are affected more frequently and more severely than females by X-linked recessive alleles. The male receives his X chromosome from his mother. Because males have just one X chromosome, all the alleles it contains are expressed, including those that cause serious and sometimes lethal medical disorders. Examples of disorders caused by X-linked recessive alleles range from relatively harmless conditions such as red-green color blindness to the always-fatal Duchenne muscular dystrophy (DMD), one of a group of muscular dystrophies characterized by the enlargement of muscles. DMD is one of the most prevalent types of muscular dystrophy and involves rapid progression of muscle degeneration early in life. It is X-linked, affects mainly males, and, according to the National Center for Biotechnology Information (2007, http://www.ncbi.nlm.nih.gov/books/bv.fcgi?call=bv.View..ShowSection&rid=gnd.section.161), 1 out of 3,500 boys worldwide are diagnosed with it. Another X-linked recessive allele causes Tay-Sachs disease, a disease that is most common among people of Jewish descent and results in neurological disorders and death in childhood.

All the male offspring of females who carry X-linked recessive alleles will be affected by the recessive allele. Female children are not as likely to express harmful recessive X-linked traits because they have two X chromosomes. Fifty percent of the female offspring will receive the recessive allele from a mother who carries the allele and an unaffected father. (See Figure 2.22.)

COMMON MISCONCEPTIONS ABOUT INHERITANCE

There are many myths and misunderstandings about genetics and inheritance. For example, some people mistakenly believe that in any population dominant traits are inevitably more common than recessive traits. This is simply untrue, as evidenced by the observation that, among humans, the allele that produces six fingers and six toes on each hand and foot, respectively, is dominant over the allele for five fingers and five toes, but the incidence of polydactyly (extra digits) is actually quite low.

Another lingering misconception is that sex-linked diseases occur only in males. This is untrue but it is easy to understand the source of the misunderstanding. For years it was thought that hemophilia (a disease characterized by uncontrolled bleeding) did not occur in females. The observation seemed reasonable because there were no reported cases of the disease among females. Even though it was true that there were no females with the disease, the reasoning was incorrect. For a female to suffer from hemophilia, she would require a defective recessive gene on both of her X chromosomes, meaning her mother was carrying the gene and the disease affected her father. Because most people with hemophilia died young, few lived to produce offspring.

TABLE 2.2

Life expectancy at selected ages, by race and gender, selected years 1900–2003

[Data are based on death certificates]

Specified age and year	All races			White			Black or African American[a]		
	Both sexes	Male	Female	Both sexes	Male	Female	Both sexes	Male	Female
	Remaining life expectancy in years								
At birth									
1900[b, c]	47.3	46.3	48.3	47.6	46.6	48.7	33.0	32.5	33.5
1950[c]	68.2	65.6	71.1	69.1	66.5	72.2	60.8	59.1	62.9
1960[c]	69.7	66.6	73.1	70.6	67.4	74.1	63.6	61.1	66.3
1970	70.8	67.1	74.7	71.7	68.0	75.6	64.1	60.0	68.3
1980	73.7	70.0	77.4	74.4	70.7	78.1	68.1	63.8	72.5
1990	75.4	71.8	78.8	76.1	72.7	79.4	69.1	64.5	73.6
1995	75.8	72.5	78.9	76.5	73.4	79.6	69.6	65.2	73.9
1996	76.1	73.1	79.1	76.8	73.9	79.7	70.2	66.1	74.2
1997	76.5	73.6	79.4	77.1	74.3	79.9	71.1	67.2	74.7
1998	76.7	73.8	79.5	77.3	74.5	80.0	71.3	67.6	74.8
1999	76.7	73.9	79.4	77.3	74.6	79.9	71.4	67.8	74.7
2000	77.0	74.3	79.7	77.6	74.9	80.1	71.9	68.3	75.2
2001	77.2	74.4	79.8	77.7	75.0	80.2	72.2	68.6	75.5
2002	77.3	74.5	79.9	77.7	75.1	80.3	72.3	68.8	75.6
2003	77.5	74.8	80.1	78.0	75.3	80.5	72.7	69.0	76.1
At 65 years									
1950[c]	13.9	12.8	15.0	—	12.8	15.1	13.9	12.9	14.9
1960[c]	14.3	12.8	15.8	14.4	12.9	15.9	13.9	12.7	15.1
1970	15.2	13.1	17.0	15.2	13.1	17.1	14.2	12.5	15.7
1980	16.4	14.1	18.3	16.5	14.2	18.4	15.1	13.0	16.8
1990	17.2	15.1	18.9	17.3	15.2	19.1	15.4	13.2	17.2
1995	17.4	15.6	18.9	17.6	15.7	19.1	15.6	13.6	17.1
1996	17.5	15.7	19.0	17.6	15.8	19.1	15.8	13.9	17.2
1997	17.7	15.9	19.2	17.8	16.0	19.3	16.1	14.2	17.6
1998	17.8	16.0	19.2	17.8	16.1	19.3	16.1	14.3	17.4
1999	17.7	16.1	19.1	17.8	16.1	19.2	16.0	14.3	17.3
2000	18.0	16.2	19.3	18.0	16.3	19.4	16.2	14.2	17.7
2001	18.1	16.4	19.4	18.2	16.5	19.5	16.4	14.4	17.9
2002	18.2	16.6	19.5	18.2	16.6	19.5	16.6	14.6	18.0
2003	18.4	16.8	19.8	18.5	16.9	19.8	17.0	14.9	18.5
At 75 years									
1980	10.4	8.8	11.5	10.4	8.8	11.5	9.7	8.3	10.7
1990	10.9	9.4	12.0	11.0	9.4	12.0	10.2	8.6	11.2
1995	11.0	9.7	11.9	11.1	9.7	12.0	10.2	8.8	11.1
1996	11.1	9.8	12.0	11.1	9.8	12.0	10.3	9.0	11.2
1997	11.2	9.9	12.1	11.2	9.9	12.1	10.7	9.3	11.5
1998	11.3	10.0	12.2	11.3	10.0	12.2	10.5	9.2	11.3
1999	11.2	10.0	12.1	11.2	10.0	12.1	10.4	9.2	11.1
2000	11.4	10.1	12.3	11.4	10.1	12.3	10.7	9.2	11.6
2001	11.5	10.2	12.4	11.5	10.2	12.3	10.8	9.3	11.7
2002	11.5	10.3	12.4	11.5	10.3	12.3	10.9	9.5	11.7
2003	11.8	10.5	12.6	11.7	10.5	12.6	11.4	9.8	12.4

—Data not available.

[a]Data shown for 1900–1960 are for the nonwhite population.

[b]Death registration area only. The death registration area increased from 10 states and the District of Columbia in 1900 to the coterminous United States in 1933.

[c]Includes deaths of persons who were not residents of the 50 states and the District of Columbia.

Notes: Populations for computing life expectancy for 1991–1999 are 1990-based postcensal estimates of U.S. resident population.
Starting with 2003 data, California, Hawaii, Idaho, Maine, Montana, New York, and Wisconsin reported multiple-race data. The multiple-race data for these states were bridged to the single race categories of the 1977 Office of Management and Budget standards for comparability with other states.

SOURCE: "Table 27. Life Expectancy at Birth, at 65 Years of Age, and at 75 Years of Age, by Race and Sex: United States, Selected Years 1900–2003," in *Health, United States, 2006*, U.S. Department of Health and Human Services, Centers for Disease Control and Prevention, National Center for Health Statistics, 2006, http://www.cdc.gov/nchs/data/hus/hus06.pdf (accessed November 21, 2006).

In other words, female hemophiliacs were rare because the pairings that might produce them were infrequent. During the 1950s the first cases of hemophilia in females were documented, and the theory was discarded.

Finally, the idea that humans are entirely unique in their genetic makeup is false. In fact, human beings share much of their genetic composition with other organisms in the natural world. Furthermore, most human genetic variation is relatively insignificant. Even variations that alter the sequence of amino acids in a protein often produce no discernable influence on the action of the protein. Differences in portions of DNA with as yet unknown functions appear to have no impact at all.

CHAPTER 3
GENETICS AND EVOLUTION

The essence of Darwinism lies in a single phrase: natural selection is the major creative force of evolutionary change. No one denies that natural selection will play a negative role in eliminating the unfit. Darwinian theories require that it create the fit as well.

—Stephen Jay Gould, "The Return of Hopeful Monsters" (*Natural History*, June–July 1977)

The term *evolution* has multiple meanings; it is most generally used to describe the theory that all organisms are linked via descent to a common ancestor. Evolution also refers to the gradual process during which change occurs. In biology it is the theory that groups of organisms, such as species, change or develop over long periods of time so that their descendants differ from their ancestors morphologically (in form, structure, and physiology) in terms of their life processes, activities, and functions. (Species are the smallest groups into which most living things that share common characteristics are divided. Among the key characteristics that define a species is that its members can breed within the group but not outside it.)

It is important to understand that not all change is considered evolution; evolution encompasses only those changes that are inheritable and may be passed on to the next generation. For example, evolution does not explain why humans are taller and bigger today than they were a century ago. This phenotypic (observable) change is attributable to changes in the environment—that is, improvements in nutrition and medicine—and is not inherited. Similarly, it should also be noted that while evolution leads to increasing complexity, it does not necessarily signify progress because an adaptation, trait, or strategy that is successful at one time may be unsuccessful at another.

In genetic terms evolution can be defined as any change in the gene pool of a population over time or changes in the frequency of alleles in populations of organisms from generation to generation. Evolution requires genetic variation, and the incremental and often uneven changes described by the process of evolution arise in response to an organism's or species' genetic response to environmental influences.

Evidence of evolution has been derived from fossil records, genetics study, and changes observed among organisms over time. The process produces the transformations that generate new species only able to survive if they can respond quickly and favorably enough to environmental changes. Population genetics is the discipline that considers variation and changing ratios of genetic types within populations to explain how populations evolve. Such changes within a population are called microevolution. In contrast, macroevolution describes larger-scale changes that produce entirely new species. Although some researchers speculate that the two processes are different, many scientists believe that macroevolutionary change is simply the final outcome of the collected effects of microevolution.

Molecular evolution is the term used to describe the period before cellular life developed on Earth. Scientists speculate that specific chemical reactions occurred that created information-containing molecules that contributed to the origin of life on this planet. Theories about molecular evolution presume that these early information-containing molecules were precursors to genetic structures capable of replication (duplication of deoxyribonucleic acid [DNA] by copying specific nucleic acid sequences) and mutation (change in DNA sequence).

NATURAL SELECTION

Natural selection is a mechanism of evolution. The principles of organic evolution by means of natural selection were described by the English naturalist Charles Darwin. Much of his early research focused on geology, and he developed theories about the origin of different land formations when he went on a five-year expedition around the world aboard the HMS *Beagle*. During his travels he developed an interest in population diversity.

When Darwin identified twelve different species of finches in the islands of the Galápagos chain off the coast of Ecuador, he speculated that the birds must have descended from a common ancestor even though they differed in terms of beak shape and overall size. The birds became known as "Darwin's finches" and are examples of a process called adaptive radiation, in which species from a common ancestor successfully adapt to their environment via natural selection. Darwin suspected that the finches had become geographically isolated from one another and after years of adapting to their distinctive environments had developed and gradually evolved into separate species incapable of interbreeding.

To explain this occurrence, Darwin relied on his own observations of the existence of variation in and between species, his knowledge of animal breeding, and the results of zoological research conducted by French naturalist Jean Baptiste Lamarck. Lamarck suggested four laws to explain how animal life might change:

- The life force tends to increase the volume of the body and to enlarge its parts.

- New organs can be produced in a body to satisfy a new need.

- Organs develop in proportion to their use.

- Changes that occur in the organs of an animal are transmitted to that animal's progeny.

Darwin famously took issue with this last point, Lamarck's theory of acquired traits, particularly his suggestion that giraffes that make their necks longer by stretching to reach the uppermost leaves on tall trees would then pass on longer necks to their offspring. However, even though Darwin discredited the specifics of Lamarck's theories concerning evolution, he agreed with Lamarck's idea that species changed over time, and he acknowledged Lamarck as an important forerunner and influence on his own work.

Darwin's ideas were also influenced by a 1798 pamphlet, written by Thomas Robert Malthus, titled *An Essay on the Principle of Population, as it Effects the Future Improvement of Society*. In this work Malthus put forth his hypothesis that unchecked population growth always exceeds the growth of the means of subsistence (the food supply needed to sustain it). In other words, if there were no outside factors stopping population growth, there would inevitably be more people than food. According to Malthus, actual population growth is kept in line with food supply growth by "positive checks," such as starvation and disease, which increase the death rate, and "preventive checks," such as postponement of marriage, which reduce the birthrate. Malthus's hypothesis suggested that actual population always tended to rise above the food supply, but that historically overpopulation had been prevented by wars, famine, and epidemics of disease.

Darwin also knew that farmers had been able to modify species of domestic animals for hundreds of years. Cattle breeders produced breeds that yielded exceptional milk production by mating their best milk producers. Superior egg-laying hens had been bred using the same technique. Because it was possible for farmers to modify a species by artificially selecting those members permitted to reproduce, Darwin hypothesized that nature might have a comparable mechanism for determining which characteristics might be passed on to future generations.

He also realized that while individual organisms in every species had the potential to produce many offspring, the natural population of any species remains relatively constant over time. Darwin concluded that the natural environment acts as a natural selector, determining over long periods of time which variations are best suited to survive and, by virtue of their survival, reproduce and pass on traits and adaptations that improve health and longevity.

Applying the principles of natural selection to the question of giraffes' neck lengths provides an explanation that is different from the one proposed by Lamarck. Less able to obtain food, short-necked giraffes faced starvation. As such, the genes linked to the potential to develop long necks were more likely to be passed to the next generation than the genes for short necks. Over time, the process of natural selection resulted in a population of giraffes with long necks.

Laboratory research and observation also refuted the theory of inheritance of acquired characteristics. When white mice had their tails cut off and were permitted to reproduce, each new generation was born with tails. Children of parents who had suffered amputations or disfiguring accidents did not share their parents' disabilities. Darwin's belief in evolution by natural selection was based on four premises:

- Individuals within a species are variable.

- Some of these variations are passed on to offspring.

- In every generation more offspring are produced than can survive.

- The survival and reproduction of individuals are not random. The individuals who survive and reproduce or reproduce the most are those with the most favorable variations. They are naturally selected.

As support for the theory of acquired inheritance diminished, appreciation of the underlying assumptions for the role of natural selection in evolution grew.

Attacks on Darwin's Theories

Darwin's theories were met with criticism from scientists and members of the clergy. Even some scientists who subscribed to evolutionary theory took issue with the concept of natural selection. Followers of Lamarck, known as Lamarckians, were among the most outspoken opponents of Darwin's theories. This was especially ironic because it was Lamarck's work that had inspired Darwin.

Other objections raised by scientists were related to how poorly inheritance was understood at that time. The notion of blending inheritance was popular. This is the idea that an organism blends together the traits it inherits from its parents. Those who endorsed it observed that, according to Darwin's assumptions, any new variation would mix with existing traits and would no longer exist after several generations. Although Gregor Mendel published a paper in 1866 proposing particulate as opposed to blended inheritance, his theory was not widely accepted until 1900, when it was revisited and confirmed by scientists.

The other objection to Darwin's theory was the argument that variation within species was limited and that, once the existing variation was exhausted, natural selection would cease abruptly. In 1907 the American biologist Thomas Hunt Morgan and his colleagues effectively dispelled this objection. Their experiments with fruit flies demonstrated that new hereditary variation occurs in every generation and in every trait of an organism.

The clergy were even more vociferous adversaries. Darwin's major works, *On the Origin of Species by Means of Natural Selection* (1859) and *The Descent of Man, and Selection in Relation to Sex* (1871), were published during a period of heightened religious fervor in England. Many religious leaders were aghast at Darwin's assertion that all life had not been created by God in one fell swoop. Moral outrage and opposition to Darwinian theory persisted into the twentieth century. Although society grew more tolerant and many religions accepted and incorporated evolutionary theory into their beliefs, in the early twenty-first century the debate was revived, as many fundamentalist Christian denominations in the United States became more vocal about their creationist beliefs.

In some instances opponents protested the teaching of evolution in schools and continued to defend creationist theory. In July 1925 a science teacher named John T. Scopes was tried in a Tennessee court for teaching his high school class Darwin's theory of evolution. Scopes had violated the Butler Act, which prohibited teaching evolution theory in public schools in Tennessee. Dubbed the "Monkey Trial" because of the simplified interpretation of Darwin's idea that humans evolved from apes, the courtroom drama pitted the defense attorney Clarence

S. Darrow against the prosecutor William Jennings Bryan in a debate that began over the teaching of evolution but became a conflict of deeply held social, intellectual, and religious values. In his acerbic account of the trial proceedings, the American social critic H. L. Mencken wrote:

> The Scopes trial, from the start, has been carried on in a manner exactly fitted to the anti-evolution law and the simian imbecility under it. There hasn't been the slightest pretense to decorum. The rustic judge, a candidate for re-election, has postured the yokels like a clown in a ten-cent side show, and almost every word he has uttered has been an undisguised appeal to their prejudices and superstitions. . . . Darrow has lost this case. It was lost long before he came to Dayton. But it seems to me that he has nevertheless performed a great public service by fighting it to a finish and in a perfectly serious way. Let no one mistake it for comedy, farcical though it may be in all its details. It serves notice on the country that Neanderthal man is organizing in these forlorn backwaters of the land, led by a fanatic, rid of sense and devoid of conscience. Tennessee, challenging him too timorously and too late, now sees its courts converted into camp meetings and its Bill of Rights made a mock of by its sworn officers of the law. There are other States that had better look to their arsenals before the Hun is at their gates.

> —"'The Monkey Trial': A Reporter's Account," http://www.law.umkc.edu/faculty/projects/ftrials/scopes/menk.htm

At the end of deliberations, Darrow requested a guilty verdict so that the case could be heard before the Tennessee Supreme Court on appeal. The jury complied, and the presiding judge fined Scopes $100. A year later, however, the Tennessee Supreme Court overturned the verdict on a technicality. Wanting to close the case once and for all, the court dismissed it altogether.

Misuse of Darwin's Theories

After Darwin's theories became well known, some made use of his terminology and concepts to argue that certain groups of human beings were naturally superior to others. The term *social Darwinism* is used to refer to these ideas, but it is important to note that Darwin himself did not believe in social Darwinism.

One example of social Darwinist thinking would be arguing that the rich and successful members of society are fitter, superior, or in some way more highly evolved than the poor. Social Darwinism has also been used to justify racism and colonialism, as in the nineteenth and early twentieth centuries, when many white Europeans and Americans asserted that they were naturally superior to Africans and Asians and used this claim to justify taking control of their land and resources. Some argued further that it was not just a right but an obligation—the "white man's burden"—for Europeans and Americans to rule over and "civilize" people in less industrialized parts of the world.

None of these social Darwinist theories are scientific in nature, and all are false. Modern genetics has shown that there is no group of human beings that is more evolved or otherwise better than the rest of humanity. Despite having been discredited by scientific research, Social Darwinism continues to be used in attempts to justify various prejudices and inequalities.

MODERN EVOLUTION-CREATION DEBATE

Arguments over the accuracy and importance of Darwin's theories have continued to the present day. Creationists believe the biblical account of the Earth's creation as it appears in the book of Genesis. Some acknowledge microevolution—changes in a species over time in response to natural selection—but they generally do not believe in speciation—that one species can beget or become another over time. There are various gradations of creationist beliefs, but all reject evolution and its argument that the interaction of natural selection and environmental factors explains the diversity of life on Earth.

At the core of the conflict is the observation that evolution threatens the view that human beings have a special place in the universe. Many creationists find it disturbing to contemplate the idea that human existence is a random occurrence, or that universal order is a chance occurrence rather than a response to a divine decree or plan.

The issue remains prominent in present-day society. In the November 2004 issue of *National Geographic* (http://magma.nationalgeographic.com/ngm/0411/feature1/fulltext.html), David Quammen asks, "Was Darwin Wrong?" Quammen, an award-winning science writer, reports that despite seemingly overwhelming evidence, many Americans believe evolution is simply an unproven speculation rather than an explanatory statement that fits the evidence. Although observation and experiment support evolutionary theory, its apparent contradiction with many religious tenets renders it unacceptable to those Americans who choose to believe that God alone, and not evolution, produced human life on Earth.

Nearly One-Half of Americans Believe Humans Did Not Evolve

Nearly a century and a half after Darwin's publication of *On the Origin of Species by Means of Natural Selection*, his theory remains highly controversial. Although scientists assert that evolution is well established by scientific evidence, Gallup Organization surveys repeatedly reveal that a substantial portion of Americans do not believe that the theory of evolution best explains the origins of human life.

For example, a 2006 Gallup Poll reveals that 46%—nearly half of the U.S. population—rejects evolution in favor of the belief that humans were created by God approximately 10,000 years ago. (See Figure 3.1.) In

FIGURE 3.1

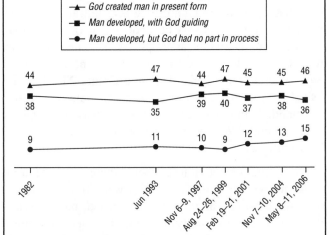

Public opinion on the origin and development of human life, selected years 1982–2006

[Numbers shown in percentages]

WHICH OF THE FOLLOWING STATEMENTS COMES CLOSEST TO YOUR VIEWS ON THE ORIGIN AND DEVELOPMENT OF HUMAN BEINGS—(HUMAN BEINGS HAVE DEVELOPED OVER MILLIONS OF YEARS FROM LESS ADVANCED FORMS OF LIFE, BUT GOD GUIDED THIS PROCESS, HUMAN BEINGS HAVE DEVELOPED OVER MILLIONS OF YEARS FROM LESS ADVANCED FORMS OF LIFE, BUT GOD HAD NO PART IN THIS PROCESS, (OR) GOD CREATED HUMAN BEINGS PRETTY MUCH IN THEIR PRESENT FORM AT ONE TIME WITHIN THE LAST 10,000 YEARS OR SO)?

SOURCE: "Which of the following statements comes closest to your views on the origin and development of human beings—(human beings have developed over millions of years from less advanced forms of life, but God guided this process, human beings have developed over millions of years from less advanced forms of life, but God had no part in this process, (or) God created human beings pretty much in their present form at one time within the last 10,000 years or so)?" in *Evolution, Creationism, Intelligent Design*, The Gallup Organization, May 2006, http://www.galluppoll.com/content/?ci=21814&pg=1 (accessed October 16, 2006). Copyright © 2006 by The Gallup Organization. Reproduced by permission of The Gallup Organization.

another study, "Science Communication: Public Acceptance of Evolution" (*Science*, August 2006), by Jon D. Miller, Eugenie C. Scott, and Shinji Okamoto, people in the United States and in thirty-two European countries were asked whether they considered the statement, "Human beings, as we know them, developed from earlier species of animals," to be true or false. Of the nations surveyed, the United States had the second-highest percentage of adults who said the statement was false and the second-lowest percentage who said the statement was true. Miller, Scott, and Okamoto conclude that "[t]he acceptance of evolution is lower in the United States than in Japan or Europe, largely because of widespread fundamentalism and the politicization of science in the United States."

Frank Newport of the Gallup Poll reports in "Almost Half of Americans Believe Humans Did Not Evolve" (June 5, 2006) and "American Beliefs: Evolution vs.

FIGURE 3.2

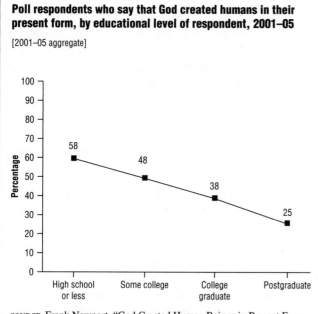

Poll respondents who say that God created humans in their present form, by educational level of respondent, 2001–05

[2001–05 aggregate]

SOURCE: Frank Newport, "God Created Human Beings in Present Form, Results by Education," in *American Beliefs: Evolution vs. Bible's Explanation of Human Origins*, The Gallup Organization, March 2006, http://www.galluppoll.com/content/?ci=21811&pg=1 (accessed October 16, 2006). Copyright © 2006 by The Gallup Organization. Reproduced by permission of The Gallup Organization.

FIGURE 3.3

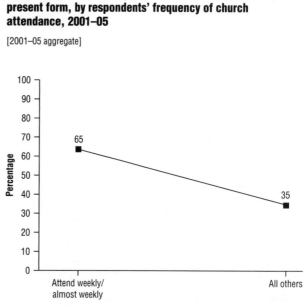

Poll respondents who say that God created humans in their present form, by respondents' frequency of church attendance, 2001–05

[2001–05 aggregate]

SOURCE: Frank Newport, "God Created Human Beings in Present Form, Results by Church Attendance," in *American Beliefs: Evolution vs. Bible's Explanation of Human Origins*, The Gallup Organization, March 2006, http://www.galluppoll.com/content/?ci=21811&pg=1 (accessed October 16, 2006). Copyright © 2006 by The Gallup Organization. Reproduced by permission of The Gallup Organization.

FIGURE 3.4

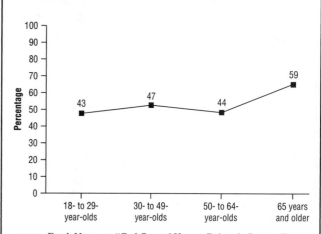

Poll respondents who say that God created humans in their present form, by age of respondent, 2001–05

[2001–05 aggregate]

SOURCE: Frank Newport, "God Created Human Beings in Present Form, Results by Age," in *American Beliefs: Evolution vs. Bible's Explanation of Human Origins*, The Gallup Organization, March 2006, http://www.galluppoll.com/content/?ci=21811&pg=1 (accessed October 16, 2006). Copyright © 2006 by The Gallup Organization. Reproduced by permission of The Gallup Organization.

Bible's Explanation of Human Origins" (March 8, 2006) that analysis of aggregate poll data from 2001 to 2005 characterizes those most likely to believe the biblical explanation of the origin of humans as those with lower levels of education, those who attend church regularly, people aged sixty-five and older, and those who identify with the Republican Party. Figure 3.2 shows that the belief that God created humans in their current form steadily decreases with advancing education. Figure 3.3 reveals that nearly twice as many Americans who attend church regularly believe the biblical explanation as do those who attend church less frequently. Figure 3.4 shows that although the proportion of Americans that accepts the biblical explanation is relatively unchanged among young and middle-aged adults, it rises sharply among adults aged sixty-five and older.

BELIEFS ABOUT EVOLUTION AMONG TEENS. A 2005 Gallup Poll asked teenagers to choose one of three statements as most consistent with their own beliefs about the origins of human life on Earth:

1. Humans developed over millions of years, but God guided the process.

2. Humans developed over millions of years, but God had no part in the process.

3. God created humans pretty much in their present form within the last 10,000 years or so.

As Figure 3.5 shows, 43% of teens believed that God guided human evolution over millions of years; more than a third of teens (38%) thought that God created humans in

FIGURE 3.5

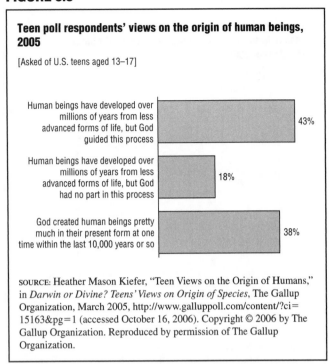

Teen poll respondents' views on the origin of human beings, 2005

[Asked of U.S. teens aged 13–17]

Human beings have developed over millions of years from less advanced forms of life, but God guided this process — 43%

Human beings have developed over millions of years from less advanced forms of life, but God had no part in this process — 18%

God created human beings pretty much in their present form at one time within the last 10,000 years or so — 38%

SOURCE: Heather Mason Kiefer, "Teen Views on the Origin of Humans," in *Darwin or Divine? Teens' Views on Origin of Species*, The Gallup Organization, March 2005, http://www.galluppoll.com/content/?ci= 15163&pg=1 (accessed October 16, 2006). Copyright © 2006 by The Gallup Organization. Reproduced by permission of The Gallup Organization.

FIGURE 3.6

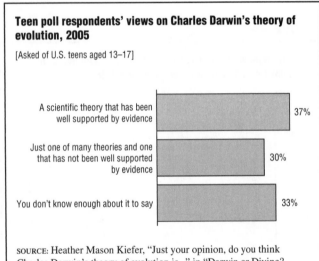

Teen poll respondents' views on Charles Darwin's theory of evolution, 2005

[Asked of U.S. teens aged 13–17]

A scientific theory that has been well supported by evidence — 37%

Just one of many theories and one that has not been well supported by evidence — 30%

You don't know enough about it to say — 33%

SOURCE: Heather Mason Kiefer, "Just your opinion, do you think Charles Darwin's theory of evolution is..." in "Darwin or Divine? Teens' Views on Origin of Species," The Gallup Organization, March 2005, http://www.galluppoll.com/content/?ci=15163&pg=1 (accessed October 16, 2006). Copyright © 2006 by The Gallup Organization. Reproduced by permission of The Gallup Organization.

more or less their present form; and just 18% believed in evolutionary theory without divine guidance.

The Gallup pollsters also asked teens whether or not they believe evolutionary theory is supported by scientific evidence. More than a third (37%) thought evolution is a theory well supported by evidence; 30% believed it is not well supported by evidence; and 33% conceded that they did not know enough to say. (See Figure 3.6.)

Some States Move to Teach Creationism in Public Schools

Because almost 50% of Americans say they do not believe in evolution, it is not surprising that the decision about which theory—evolution or creationism—should be taught in public schools has been hotly contested. The American Institute of Biological Sciences (http://www .aibs.org/public-policy/evolution_state_news.html), a not-for-profit organization dedicated to advancing biological research and education, reports that during 2006 there were school board battles over the issue in twenty-seven states.

Some opponents of evolution propose an alternative known as intelligent design (ID), an explanation that credits intelligence, rather than an undirected process such as natural selection, as the source of life on Earth. Proponents of ID disagree with a basic tenet of evolutionary theory: Darwin's claim that the complex design of biological systems resulted by chance. They contend that direction from an intelligent designer—a supernatural being—is necessary to explain adequately the origins and complexity of life on Earth, particularly human life.

Throughout the United States, school boards have considered whether they want to teach students Darwin's theory, creationism, or alternatives such as ID. In January 2005, eighty years after the Scopes trial, the school board in rural Dover, Pennsylvania, became the first in the nation to require that students be taught that an alternative to evolution exists. A 2005 Gallup Poll found Americans divided about which explanation of the origin of human life should be taught in public school science classes. Table 3.1 shows that 61% believed evolution should be taught, 54% thought creationism should be taught, and 43% wanted ID taught in public school science classes. As of November 2006, no state school board had mandated the teaching of ID exclusively; however, many school boards have moved to include it in the science curricula.

TABLE 3.1

Public opinion about which explanation of the origin of human life should be taught in public school science classes, 2005

ON A DIFFERENT SUBJECT, DO YOU THINK EACH OF THE FOLLOWING EXPLANATIONS ABOUT THE ORIGIN AND DEVELOPMENT OF LIFE ON EARTH SHOULD OR SHOULD NOT BE TAUGHT IN PUBLIC SCHOOL SCIENCE CLASSES, OR ARE YOU UNSURE?

2005 Aug 8–11 (sorted by "yes, should")	Yes, should	No, should not	Unsure	No answer
Evolution	61%	20	19	*
Creationism	54%	22	23	1
Intelligent design	43%	21	35	1

SOURCE: "On a different subject, do you think each of the following explanations about the origin and development of life on earth should or should not be taught in public school science classes, or are you unsure? How about—[RANDOM ORDER]?" in *Education*, The Gallup Organization, October 2005, http://www.galluppoll.com/content/?ci=1612& pg=1 (accessed October 16, 2006). Copyright © 2006 by The Gallup Organization. Reproduced by permission of The Gallup Organization.

Even some staunch advocates of Darwin's theory do not necessarily wish to exclude the teaching of creationism or ID—they would simply prefer that these alternative explanations be taught in the context of religion classes rather than in science classes. There are even those who advocate both theories. The Vatican has stated that it does not consider evolution to be in conflict with Christian faith. Francis Collins, a committed Christian and director of the Human Genome Institute at the National Institutes of Health, has repeatedly expressed his view that God created the universe and chose the remarkable mechanism of evolution to create plants, animals, and humans.

VARIATION AND ADAPTATION

Effective adaptations and variations are perpetuated in a species and tend to be incorporated into the normal or predominant phenotype for most individuals in the species. Variation persists, but it ranges around an evolutionarily determined norm. This is called adaptive radiation. For example, over time Darwin's finches have developed beaks best suited to their functions. The finches that eat grubs have long, thin beaks to enter holes in the ground and pull out the grubs. Finches that eat buds and fruit have clawlike beaks to grind their food, giving them a survival advantage in environs where buds are the only available food source. In another example of adaptive radiation, present-day giraffes have necks of varying lengths, but most tend to be long.

Ancient humans underwent many evolutionary changes. An example of adaptive radiation in humans is the development and refinement of the upper limbs to perform fine motor skills necessary to make and use complex tools. Another adaptation is the quantity of the pigment melanin present in the skin. People from areas near the equator have more melanin, an adaptation that darkens their skin and protects them from the sun. Anatomical structure also appears to have responded to the environment. Those who thrive in colder climates produce offspring that are shorter and broader than those who live in warmer climates. This adaptive stature, with its relatively low surface area, enables those people to conserve rather than lose body heat. For example, native Alaskans, who are generally of short stature, are well suited to their cold climate.

Natural selection does not produce uniformity or perfection. Instead, it generates variability that persists when it helps a species to adapt to, and thrive in, its environment. It acts on outward appearance (phenotypes), not on internal coding (genotypes), and it is not a process that always discards individual genes in favor of others that might produce traits better suited for survival. For traits attributable to multiple genes, many different combinations of gene pairs may produce the same or comparable phenotypes. Multiple phenotypes may be neutral or even

beneficial in terms of survival in a given environment, and there is no reason for such variation to be eliminated by natural selection.

Furthermore, natural selection does not completely or rapidly eliminate genes that produce traits unsuited for adaptation or survival. Even though some individuals with harmful traits die young or do not reproduce, some do reproduce and pass their genes and traits to the next generation. Culling out these genes may require several generations. In other instances, seemingly harmful genes may be retained in the gene pool because there may be circumstances or environments in which their presence would improve survival. For example, the recessive allele that causes sickle-cell diseases (sickle-cell anemia and sickle B-thalassemia, in which the red blood cells contain abnormal hemoglobin) may have had a role in survival in some parts of the world. Although people who are pure recessive for this trait become ill and die prematurely, those who are hybrid for the trait may have retained a survival advantage in areas where malaria is present because people who have the sickle-cell trait or anemia are immune to the effects of malaria. This phenomenon is called balanced polymorphism and is yet another example of the seemingly counterintuitive actions of natural selection. Even though the sickle-cell trait is not beneficial on its own, historically it was advantageous in areas where malaria was a greater threat to survival than sickle-cell diseases.

By definition, natural selection is an unending, continuous process. The popular understanding of natural selection as "survival of the fittest" is somewhat misleading because organisms and species with phenotypes most suited to survive in their environments are not necessarily the "fittest." Present-day examples of natural selection include the evolution of bacteria that are antibiotic resistant and insects that resist extermination with pesticides.

Increasing and Decreasing Genetic Variation

A gene pool comprises the alleles for all the genes in a population. Gene pools in natural populations contain considerable variation, and for evolution to proceed there must be mechanisms to create and increase genetic variation. Mutation, a change in a gene, serves to create or increase genetic variation. Recombination, a process that creates new alleles and new combinations of alleles, also increases genetic variation. Another way for new alleles to enter a gene pool is by migrating from another population. In closely related species new organisms can enter a population, mate within it, and produce fertile hybrids. This action is known as gene flow.

The tendency toward increased genetic variation within a population is balanced by other mechanisms that act to decrease it. For example, some variations have the effect of limiting an organism's ability to reproduce. Any such variations, as well as those incidentally paired with them, will be nonadaptive because they will have a lower probability of being passed to the next generation. Under such circumstances, the process of natural selection serves to reduce genetic variation. In fact, differences in reproductive capability are often called natural selection. Whenever natural selection weeds out nonadaptive alleles by limiting an organism's reproductive capability, it is depleting genetic variation within the population.

Genetic Drift

Other evolutionary mechanisms contribute to genetic variation. Genetic drift is random change in the genetic composition of a population. It may occur when two groups of a species are separated and as a result cannot reproduce with each other. The gene pool of these groups will naturally differ over time. When the two groups are reunited and reproduce, gene migration occurs as their genetic differences combine, serving to increase genetic diversity. If they remain separate, their genetic differences may become so great that they develop into two separate species.

In small populations genetic drift can cause relatively rapid change because each individual's alleles constitute a large proportion of the gene pool and when an individual does not reproduce, the results are felt more acutely in smaller, rather than larger, populations. When small populations are affected by genetic drift, they may suffer a loss of valuable diversity. For this reason scientists and others involved in conservation, such as zoo curators of endangered species, make every effort to ensure that populations are large enough to withstand the effects of genetic drift.

MUTATION

Variation is the essence of life, and mutations are the source of all genetic variation. A staggering number and variety of alterations, rearrangements, and duplications of genetic material have occurred since the first living cells, in which there is an incredible range of life forms, from amoebas and fruit flies to giant dinosaurs and humans. These dramatically different life forms were all produced using genetic material that was present and reproduced from the first living cells. They resulted from the process of mutation, an alteration in genetic material.

Perhaps because of their negative depiction in science fiction and horror films, there is a widespread common misperception that all mutations are harmful, dramatic, and deleterious. Most mutations are not harmful, and the same limited number of mutations has probably been recurring in each species for millions of years. Many helpful mutations have already been incorporated into the normal genotype through natural selection.

Harmful mutations do occur, and while they tend to be eliminated through natural selection, they do recur randomly. It is a mistake, however, to consider mutation as sudden or exclusively harmful. Mutation cannot generate sudden, drastic changes—it requires many generations to generalize throughout a species population. Furthermore, if the changes caused by mutation did not favor survival, natural selection would work against their generalization throughout the species population. In the absence of mutation, there would have been no development of life and no evolution would have occurred.

The impact of mutation varies greatly. Though mutations are changes in genetic material, they do not necessarily affect an individual organism's phenotype. When phenotype is affected, it is because the code for protein synthesis has been changed. Whether mutation will affect phenotype and the extent to which it will be influenced depends on how protein manufacture is affected, when and where the mutation occurs, and the complexity of the genetic controls governing the selected trait.

When a trait is governed by the interaction of many genes, with each exerting about the same influence, the impact of a mutation might be negligible. In traits controlled by a single pair of genes, mutation is likely to exert a much greater influence. For traits governed by the interaction between a major controlling gene pair and several other less influential gene pairs, the location of the mutation will determine its impact. Other considerations such as whether the mutated gene is dominant or recessive also determine whether it will directly act on an individual's phenotype—a mutated recessive gene might not appear in the phenotype for several generations.

Mutation can occur in any cell in an organism's body, but only germinal mutations (those that affect the cells that give rise to sperm or eggs) are passed to the next generation. When mutation occurs in somatic (other than sperm and egg) cells in the body such as muscle, liver, or brain cells, only cells that derive from mitotic division of the affected cell will contain the mutation. Although mutations in somatic cells can cause disease, most do not have a significant impact, because if the mutated gene is not involved in the specialized function of the affected cell, these "silent mutations" will not be detected. Furthermore, mutations that appear only in somatic cells disappear when the organism dies; they are not passed on to subsequent generations and do not enter the gene pool that is the source of genetic variation for the species.

Another reason that many mutations are not expressed in the phenotype is that they affect only one copy of the gene, leaving diploid organisms with an intact copy of the gene. These types of mutations have a recessive inheritance pattern and do not affect phenotype unless an individual inherits two copies of the mutation. There is also the question of the probability of a mutation in a gamete affecting offspring. The mutation may occur in a single gamete and so may only be passed on if that particular gamete is involved in conception. For example, human semen contains more than 50 million sperm per ejaculate, so it is unlikely that a mutation carried by a single sperm will be passed on. When mutation occurs during embryonic (before birth) development and all gametes are affected, there is a greater chance that it might influence the phenotype of future generations. Alternatively, if, like many mutations, the gene defect occurs with advancing age, and the affected individuals are beyond their reproductive years, then it will have no impact on future generations.

How Different Types of Genetic Mutations Occur

Mutation is a normal and fairly frequent occurrence, and the opportunity for a mutation to take place exists every time a cell replicates. In general the cells that divide many times throughout the course of an organism's life—such as skin cells, bone marrow cells, and the cells that line the intestines—are at greater risk for mutation than those that divide less frequently, such as adult brain and muscle cells.

Although DNA nearly always reproduces itself accurately, even a minor alteration produces a mutation that may alter a protein, prevent its production, or have no effect at all. There are five broad classes, and within them many varieties, of mutations, and each is named for the error or action that causes it:

- Point mutations are substitutions, deletions, or insertions in the sequence of DNA bases in a gene. The most common point mutation in mammals is called a base substitution and occurs when an A-T pair replaces a G-C pair. Base substitutions are further classified as either transitions or transversions. Transitions occur when one pyrimidine (C or T) is substituted for the other and one purine (A or G) is substituted on the other strand of DNA. Transversions occur when a purine replaces a pyrimidine. Sickle-cell anemia results from a transversion in which T replaces A in the gene for a component of hemoglobin.

- Structural chromosomal aberrations occur when the DNA in chromosomes is broken. The broken ends may remain loose or join those occurring at another break to form new combinations of genes. When movement of a chromosome section from one chromosome to another takes place, it is called translocation. Translocation between human chromosomes 8 and 21 has been implicated in the development of a specific type of leukemia (cancer of the white blood cells). It has also been shown to cause infertility (inability to sexually reproduce) by hindering the distribution of chromosomes during meiosis.

- Numerical chromosomal aberrations are changes in the number of chromosomes. In a duplication mutation genes are copied, so the new chromosome contains all of its original genes plus the duplicated one. Polyploidy is a numerical chromosomal aberration in which the entire genome has been duplicated and an individual who is normally diploid (having two of each chromosome) becomes tetraploid (containing four of each chromosome). Polyploidy is responsible for the creation of thousands of new species, acting to increase genetic diversity and produce species that are bigger, stronger, and more able to resist disease.

- Aneuploidy refers to occasions when just one or a few chromosomes are involved and describes the loss of a chromosome. Examples of aneuploidy are Down syndrome (multiple, characteristic physical and cognitive disabilities), in which there is an extra chromosome 21—usually caused by an error in cell division called nondisjunction—and Turner's syndrome, in which there is only one X chromosome. Common characteristics of Turner's syndrome include short stature and lack of ovarian development as well as increased risk of cardiovascular problems, kidney and thyroid problems, skeletal disorders such as scoliosis (curvature of the spine) or dislocated hips, and hearing and ear disturbances.

- Transposon-induced mutations involve sections of DNA that copy and insert themselves into new locations on the genome. Transposons usually disrupt and inactivate gene function. In humans selected types of hemophilia have been linked to transposon-induced mutations.

Frequency and Causes of Mutation

In the absence of external environmental influences, mutations occur rarely and are seldom expressed because many forms of mutation are expressed by a recessive allele. The most common naturally occurring mutations arise simply as accidents. Susceptibility to mutation varies during the life cycle of an organism. For example, among humans mutation of egg cells increases with advancing age—the older the mother the more likely she is to carry gametes with mutations. Susceptibility also varies among members of a species such as humans based on their geographic location and ethnic origin.

The mutation rate is the frequency of new mutations per generation in an organism or a species. Mutation rates vary widely from one gene to another within an organism and between organisms. The mutation rate for bacteria is 1 per 100 million genes per generation. Despite this relatively low rate, the enormous number of bacteria—there are more than twenty billion produced in the human intestines each day—translates into millions of new mutations to the bacteria population every day. The human mutation rate is estimated at 1 per 10,000 genes per generation, and human mutation rates are comparable throughout the world. The only exceptions are populations that have been exposed to factors known as mutagens that cause a change in DNA structure and as a result increase the mutation rate. With approximately 25,000 genes, a typical human contains three mutations. Considering the fact that human genes mutate approximately once every 30,000 to 50,000 times they are duplicated, and in view of the complexity of the gene replication process, it is surprising that so few "mistakes" are made.

Gene size, gene base composition, and the organism's capacity to repair DNA damage are closely linked to how many mutations occur and remain in the genome. Larger genes are more susceptible than smaller ones because there are more opportunities and potential sites for mutation. Organisms better able to repair and restore DNA sequences, such as many kinds of yeast and bacteria, will be less likely to have high mutation rates.

The overwhelming majority of human mutations arise in the father, as opposed to the mother. When compared to egg cells, there are more opportunities for mutation during the many cell divisions needed to produce sperm. Sperm are produced late in the life of males, whereas females produce eggs earlier—during embryonic development and are born with their full complement of eggs. Thus, the frequency of mutations that are passed to the next generation increases with parental age.

The mutation rate is partly under genetic control and is strongly influenced by exposure to environmental mutagens. Some mutagens act directly to alter DNA, and others act indirectly, by triggering chemical reactions in the cell that result in breakage of a gene or group of genes. Radiation or ultraviolet light—from natural sources such as the sun or radioactive material in the earth or from manmade sources such as X-rays—can significantly accelerate the rate of mutations.

The link between radiation and mutation was first identified in the 1920s by Hermann Joseph Muller, who was a student of Thomas Hunt Morgan and worked in the famous "Fly Room." Muller discovered that he could increase the mutation rate of fruit flies more than a hundredfold by exposing reproductive cells to high doses of radiation. His pioneering work prompted medical and dental professionals to exercise caution and minimize patients' exposure to radiation. Because mutagens in the reproductive cells are likely to affect heredity, special precautions such as donning a lead apron to block exposure are used when dental X-rays or other diagnostic imaging studies are performed. Similarly, special care is taken to prevent radiation exposure to the embryo or fetus because a mutation during this period of development when cells are rapidly proliferating might be incorporated in many cells and could result in birth defects.

MODERN SYNTHESIS OF EVOLUTIONARY GENETICS

Present-day theories of evolutionary genetics are indebted to Darwin for his groundbreaking descriptions of organisms, individuals, and speciation. Modern theory differs considerably in that it addresses evolutionary mechanisms at the level of populations, genes, and phenotypes and incorporates understanding of actions, such as genetic drift, that Darwin had not considered.

The modern synthesis of evolutionary theory differs from Darwinism by identifying mechanisms of evolution that act in concert with natural selection, such as random genetic drift. It asserts that characteristics are inherited as discrete entities called genes and that variation within a population results from the presence of multiple alleles of a gene. The most controversial tenet of modern evolutionary theory is its contention that speciation is usually the result of small, gradual, and incremental genetic changes. In other words, macroevolution is simply the cumulative effect of microevolution. Some evolutionary biologists have instead embraced the theory of punctuated equilibrium set forth by the paleontologists Niles Eldredge and Stephen Jay Gould in 1972.

Punctuated equilibrium suggests that long periods of stasis (stability) are followed by rapid speciation. It posits that new species arose rapidly during a period of a few thousand years and then remained essentially unchanged for millions of years before the next period of adaptation. Punctuated equilibrium also proposes that change occurred in a small portion of the population, rather than uniformly throughout the population. Although it is different from Darwin's theory of speciation, it is not inconsistent with natural selection; it simply presumes different mechanisms and timetables for the development of new species.

CHAPTER 4
GENETICS AND THE ENVIRONMENT

[N]ature prevails enormously over nurture when the differences in nurture do not exceed what is commonly found among persons of the same rank in society and in the same country.

—Francis Galton, *Memories of My Life* (1908)

Although genetics clearly controls many of an organism's traits, it is simplistic and incorrect to assume that organisms, including humans, are completely defined by their genes. Even though there are some phenotypes (observable traits) that are exclusively controlled by either genetics or environment, most are influenced by a complex interaction of the two. Modern genetics study defines environment as every influence other than genetic—such as air, water, diet, radiation, and exposure to infection—and it subscribes to the overarching assumption that every trait of every organism is the product of some set of interactions between genes and the environment.

NATURE VERSUS NURTURE

The debate over the relative contributions of genetics and environment—nature versus nurture—remains unresolved in many fields of study, from education to animal behavior to human disease. Historically, scientists assumed opposing viewpoints, choosing to favor either nature or nurture, rather than exploring the ways in which both play critical and complementary roles. Proponents of nature, or a person's inborn traits, argued that human characteristics are uniquely and primarily conditioned by genetics, disputing the view of those who asserted that environmental influences and experiences (nurture) determined differences between individuals and populations.

Genes versus Environment

When researchers analyze the origins of disease, the terms used to describe causation are *genetic* versus *environmental*, but the issues are the same as those in the nature-versus-nurture debate. Conditions considered to be primarily genetic are ones in which the presence or absence of genetic mutations determines whether an individual or population will develop a disease, independent of environmental exposures or circumstances. A disease considered to be primarily environmental is one in which people of virtually any genetic background can develop the disease when they are exposed to the specific environmental factors that cause it.

Even the conditions and diseases once believed to be at either end of the continuum—caused by either purely genetic or purely environmental factors—may not be exclusively attributable to one or the other. For example, an automobile accident that results in an injury might be deemed entirely environmentally caused, but many geneticists would contend that risk-taking behaviors such as the propensity to exceed the speed limit are probably genetically mediated. Furthermore, the course and duration of rehabilitation and recovery from an injury or illness is also likely genetically influenced.

At the other end of the continuum are diseases believed to be predominantly genetic in origin, such as sickle-cell anemia. Even though this disease does not have an environmental cause, there are environmental triggers that may determine when and how seriously the disease will strike. For example, sickle-cell attacks are more likely when the body has an insufficient supply of oxygen, so people with the sickle-cell trait who live at high altitudes or those who engage in intense aerobic exercise may be at increased risk of attacks. There are many more conditions for which the risk of developing the disease is strongly influenced by both genetic and environmental factors. Multiple genes and environmental factors may be involved in causing a given condition and its expression.

Asthma: A Disease with Genetic and Environmental Causes

Asthma, a disorder of the lungs and airways that causes wheezing and other breathing problems, is a good example of how genes and the environment can interact to cause diseases. The scientific evidence that asthma had a genetic basis came from studies conducted between the 1970s and 1990s, which found patterns of inheritance in families and genetic factors involved in the severity and triggers for asthma attacks. In "Family Concordance of IgE, Atopy, and Disease" (*Journal of Allergy and Clinical Immunology*, February 1984), Michael D. Lebowitz, Robert Barbee, and Belton Burrows looked at 350 families and found that in those where neither parent had asthma, 6% of the children had asthma; when one parent had asthma, 20% were affected; and when both parents had the condition, 60% of their offspring were affected.

In "Genetic Factors in the Presence, Severity, and Triggers of Asthma" (*Archives of Disease in Childhood*, August 1995), a study of pairs of twins in which at least one twin had asthma, Edward P. Sarafino and Jarrett Goldfedder discovered that genetics and environment both made strong contributions to the development of the illness. If asthma was directed solely by genes, then 100% of the identical twins, who are exactly the same genetically, would be expected to have asthma (the concordance rate is the rate of agreement, when both members of the pair of twins have the same trait). Instead, Sarafino and Goldfedder found that just over half (59%) of twins both had asthma. If asthma was entirely environmentally caused, then genes should make no difference at all—the concordance rate would be the same for identical and fraternal (nonidentical) twins. Sarafino and Goldfedder found that the concordance rate of 59% was more than twice as high in identical twins as in fraternal twins (24%). This research offers clear confirmation that asthma has a significant genetic component.

Asthma is an example of a condition that is polygenic—controlled by more than one gene. Familial studies demonstrate that asthma does not conform to simple Mendelian patterns of inheritance and that multiple independent segregating genes are required for phenotypic expression (polygenic inheritance). Nevertheless, it has long been known that the complex phenotypes of asthma have a significant genetic contribution. Researchers speculate that several genes combine to increase susceptibility to asthma and that there are also genes that lower susceptibility to developing the condition. Genes also affect asthma severity and the way people with asthma respond to various medications.

Environmental triggers for asthma include allergens such as air pollution, tobacco smoke, dust, and animal dander. Other environmental factors linked to its development are a diet high in salt, a history of lung infections, and the lack of siblings living at home. Researchers speculate that younger children in families with older siblings are less likely to develop asthma because their early exposures to foreign substances (such as dirt and germs) brought into the home by older siblings heighten their immune systems. The enhanced immune response is believed to have a protective effect against asthma and other illnesses.

Harvard Medical School researchers Michael E. Wechsler and Elliot Israel, in "The Genetics of Asthma" (*Seminars in Respiratory and Critical Care Medicine*, 2002), acknowledge some of the difficulties faced by researchers studying a complex disease such as asthma. They describe the following challenges faced by geneticists seeking to pinpoint the genetics of asthma:

- Population studies suggest that asthma is a polygenic disease, with many diverse locations of possible asthma genes already identified.

- The definition of asthma can vary from one health care practitioner to another. Some make the diagnosis based on changes in airway reactivity, others on levels of airway function, and still others on clinical symptoms. As a result, phenotyping methods must be examined and carefully compared because the variety of clinical symptoms a patient with asthma may present, such as cough, shortness of breath, wheezing, and chest tightness, are also common to several other conditions such as bronchitis or heart failure, and confusion may result in misdiagnosis.

- Even though a clinical history of asthma symptoms or phenotypes often suggests a diagnosis of asthma, there is no definitive, specific definition that classifies an individual as having or not having asthma. As a result, some people may be incorrectly labeled or identified, potentially yielding false data or nonreplicable results.

- Several recent epidemiological (the study of the spread of disease in a population) studies suggest that asthma may have many different phenotypic expressions at different ages as assessed by their risk factors and prognosis. For example, children under age six who at various times experience wheezing are labeled as asthmatic. However, most of these children do not continue to have asthma symptoms as they age—they seem to outgrow the condition.

- Asthma differs in terms of severity (mild, moderate, or severe and intermittent or persistent), suggesting different genetic or environmental influences and triggers. There are also several different subgroups of asthma patients, including aspirin-sensitive asthmatics and exercise-induced asthmatics. Each of these variations may have a different biological mechanism that accounts for each individual's phenotype.

Wechsler and Israel conclude that while a small percentage of cases of asthma may result from a single gene defect or a single environmental factor, asthma is a complex genetic disease that cannot be explained by single-gene models. In most instances it appears to result from the interaction of multiple genetic and environmental factors. Similar to other complex diseases such as diabetes and hypertension (high blood pressure), the complexity of asthma genetics may be characterized by the contribution of different genes and different environments in different populations. Population studies of asthma face challenges comparable to genetics studies of other common complex traits. They are complicated by genetic factors such as incomplete penetrance (transmission of disease genes without the appearance of the disease), genetic heterogeneity (mutations in any one of several genes that may result in similar phenotypes), epistasis (when the effects of multiple genes have a greater effect on phenotype than individual effects of single genes), polygenic inheritance (mutations in multiple genes simultaneously producing the affected phenotype), and gene-environment interactions.

Genetic Susceptibility

Genetic susceptibility is the concept that the genes an individual inherits affect how likely he or she is to develop a particular condition or disease. When an individual is genetically susceptible to a particular disease, his or her risk of developing the disease is higher. Genetic susceptibility interacts with environmental factors to produce disease, but genes and environment do not necessarily make equal contributions to causation. Genes can cause a slight or a strong susceptibility. When the genetic contribution is weak, the environmental influence must be strong to produce disease, and vice versa.

In most instances a susceptibility gene strongly influences the risk of developing a disease only in response to a specific environmental exposure. If the environmental exposure occurs infrequently, the gene will be of low penetrance, and it may seem that the environmental exposure is the primary cause of the disease, even though the gene is required for developing the disease. For this reason, even when environmental agents are suspected to be a major cause of a particular disease, there is still the possibility that genetic factors also play a major part, particularly genetic mutations with low penetrance. Similarly, a critical mix of nature and nurture is likely to determine individual traits and characteristics. Genetic factors may be considered as the foundation on which environmental agents exert their influence. Based on this premise, it is now widely accepted that while certain environmental factors alone and certain genetic factors alone may explain the origins of some traits and diseases, most of the time the interaction of both genetic and environmental factors will be required for their expression.

Since the 1990s researchers have identified more and more genes that influence an individual's susceptibility to

disease. Scientists have already linked specific deoxyribonucleic acid (DNA) variations with increased risk of common diseases and conditions, including cancer, diabetes, hypertension, and Alzheimer's disease (a progressive neurological disease that causes impaired thinking, memory, and behavior). The questions that persist in the genes-versus-environment debate no longer focus on whether a particular trait or disease is caused exclusively by a specific gene. Instead, researchers continue to explore the extent to which genes and environment influence the development of specific traits, especially conditions linked to health and susceptibility to disease.

Examples of the kinds of questions about susceptibility that geneticists and other medical researchers hope to answer include:

- Why do some tobacco smokers live long, healthy lives while others develop lung cancer and die early?

- Why do some people who are repeatedly exposed to the human immunodeficiency virus, the virus that causes acquired immune deficiency syndrome, resist contracting the virus?

- Why do common allergens (substances that cause allergies) cause moderate discomfort for some people and life-threatening asthma for others?

TWIN STUDIES

The English anthropologist Francis Galton, a cousin of Charles Darwin, conducted some of the first reported twin studies. Advancing his cousin's theories of evolution, he performed twin studies in 1876 to investigate the extent to which the similarity of twins changes over the course of development. He is considered the originator of the field of medical genetics because of his considerable contributions to the nature-versus-nurture debate and the research methods he developed for evaluating heritability. Genetic determination is the combination of genes that creates a trait or characteristic, and heritability is what causes differences in those characteristics.

Twin studies, meaning comparisons of identical (monozygotic) twins to fraternal or nonidentical (dizygotic) twins, are performed to estimate the relative contributions of genes and environment—that is, the extent to which environmental-versus-genetic influences operate on specific traits. Twin studies also help to determine the proportion of the variability in a trait that might be because of genetic factors. The studies aim to identify the causes of familial resemblance by comparing the concordance rates of monozygotic twins and dizygotic twins. Monozygotic twins share the same genetic material—they have 100% of their genes in common—and dizygotic twins share only half their genetic material—they have 50% of their genes in common. Most twin studies report their results in terms of pairwise concordance rates, which measure the number of pairs, or probandwise concordance rates, which count the number of individuals.

Monozygotic twins serve as excellent subjects for controlled experiments because they share prenatal environments and those reared together also share common family, social, and cultural environments. Furthermore, studies of twins can both point to hereditary effects and estimate heritability, a term that describes the magnitude of the genetic effect. The limitations of such studies include the potential to overestimate or underestimate the role of genetics if environmental influences treat twins as more alike or more different than they actually may be. In addition, in some studies it has been difficult to control for other potential causes or sources of variation.

Some of the most conclusive twin study research has analyzed identical and fraternal twins who were raised apart. Researchers have sought to establish whether characteristics such as personality traits, aptitudes, and occupational preferences are the products of nature or nurture. Similar characteristics among identical twins reared apart might indicate that their genes played a major role in developing that trait. Different characteristics might indicate the opposite—that environmental influences assume a much stronger role. By comparing monozygotic and dizygotic twins, investigators can test their hypotheses and confirm the findings of earlier research. For example, if identical twins raised in different homes have many similarities, but fraternal twins raised apart have little in common, researchers may conclude that genes are more important than environment in determining specific characteristics, traits, susceptibilities, and diseases.

Paul Lichtenstein et al., in "Environmental and Heritable Factors in the Causation of Cancer—Analyses of Cohorts of Twins from Sweden, Denmark, and Finland" (New England Journal of Medicine, July 13, 2000), looked at nearly 45,000 pairs of twins to determine if the likelihood of developing certain kinds of cancers is more closely linked to environmental exposures than genetics. The results of this large-scale study revealed that environmental factors are linked to twice as many cancers as genetic factors. In fact, the risk of developing only three types of cancer (albeit some of the most common cancers)—breast, colorectal, and prostate cancers—show a significant genetic correlation.

Prostate cancer is found to have the strongest genetic link, with 42% of risk explained by genetic factors and 58% by environmental factors. The other cancers with a demonstrable genetic link, breast and colorectal cancers, are found to have less than a 35% link to genetics. Lichtenstein and his colleagues conclude that inherited genetic factors make a minor contribution to susceptibility to most types of cancer.

Do Genes Govern Sexual Orientation?

The role of genetics in establishing sexual orientation (the degree of sexual attraction to men or women) and its link to homosexuality have been hotly debated in the relevant scientific literature and the media. Studies of identical twins reveal that sexual orientation, like the overwhelming majority of human traits and characteristics, is not exclusively governed by genetics, but is more likely the result of a gene-environment interaction. For example, if homosexuality was exclusively controlled by genes, then either both members of a set of identical twins would be homosexual or neither would be. Multiple studies show that if one twin is homosexual his or her sibling is also homosexual less than 40% of the time.

J. Michael Bailey, Michael P. Dunne, and Nicholas G. Martin systematically evaluated gender identity and sexual orientation of twins and reported their findings in "Genetic and Environmental Influences on Sexual Orientation and Its Correlates in an Australian Twin Sample" (Journal of Personality and Social Psychology, March 2000). Bailey, Dunne, and Martin observed that both male and female homosexuality appears to run in families and that studies of unseparated twins suggest that this is primarily because of genetic rather than familial environmental influences. They also observe that previous research suffers from limitations such as recruiting subjects via publications aimed at homosexuals or by word of mouth—strategies likely to bias the samples and results.

To overcome these limitations, Bailey, Dunne, and Martin assessed twins from the Australian Twin Registry rather than recruiting twins especially for the purpose of their research. Using probandwise concordance (an estimate of the probability that a twin is nonheterosexual given that his or her co-twin is nonheterosexual), they found lower rates of twin concordance for nonheterosexual orientation than in previous studies. The most striking difference was between the researchers' probandwise concordance rates and those of past twin studies of sexual orientation. Previously, the lowest concordances for single-sex identical twins were 47% for women and 48% for men. This study documents concordances of just 20% for women and 24% for men, significantly lower than the rates reported for the two largest previous twin studies of sexual orientation. Bailey, Dunne, and Martin conclude that sexual orientation is familial; however, their study does not provide statistically significant support for the importance of genetic factors for this trait. They caution that this does not mean that their results entirely exclude heritability. In fact, they consider their findings consistent with moderate heritability for male and female sexual orientation, even though their male monozygotic concordance suggests that any major gene for homosexuality has either low penetrance or low frequency.

Bailey, Dunne, and Martin attribute their markedly different results to the observation that in previous studies twins deciding whether to participate in research that was clearly designed to study homosexuality probably considered the sexual orientation of their co-twins before agreeing to participate. In contrast, the more general focus of the Bailey, Dunne, and Martin study and its anonymous response format made such considerations less likely.

Even though it remains unclear from recent studies whether concordance is closer to 50% or 30%, all researchers concur that it is not 100%. This finding suggests that the influence of genes on sexual orientation is indirect and influenced by environment. Neil Whitehead and Briar Whitehead claim in *My Genes Made Me Do It* (1999) that "genes make proteins, not preferences." Similarly, they contend that "genes create a tendency, rather than a tyranny" and conclude that all the identical twin studies reveal that in terms of determining sexual orientation, neither genetic nor family-related factors are overwhelming. Furthermore, Whitehead and Whitehead believe that all influences—genetic and environmental—are subject to change and that it is possible to "foster or foil genetic or family influences."

IS THERE A GAY GENE? In "New Genetic Regions Associated with Male Sexual Orientation Found" (*WebMD Medical News Archive*, January 28, 2005), Jennifer Warner described research that examined the entire human genetic makeup—genetic information on all chromosomes—in an effort to identify possible genetic determinants of male sexual orientation. Investigators looked at the genetic makeup of 456 men from 146 families with two or more gay brothers and found the same genetic patterns among the gay men on three chromosomes: 7, 8, and 10. Sixty percent of the gay men in the study shared these common genetic patterns, which was slightly more than the 50% expected by chance alone. Patterns involving chromosomes 7 and 8 were associated with sexual orientation regardless of whether the man received them from his father or his mother; however, the areas on chromosome 10 were only associated with male sexual orientation if they were inherited from the mother. The identification of these regions has spurred further research to identify the individual genes in these regions that are linked to sexual orientation.

GENETIC AND ENVIRONMENTAL INFLUENCES ON INTELLIGENCE

The role of genetics in determining a person's intelligence is a controversial subject. Few would deny that genes play some role, but many are uncomfortable with the idea that genes determine intelligence. For if intelligence is a genetic trait, the implication is that some people are born to be smart, others are not, and education and upbringing cannot change it. The results of many studies and contentious debate in the scientific community have produced little consensus about the relationship between genetics and intelligence. At least part of the problem stems from the fact that the term *intelligence* is defined differently by different people.

Although this issue has been argued since the 1870s—when Galton proposed his controversial and arguably racist notions about the heritability of intelligence—the debate was reignited during the 1990s when Richard J. Herrnstein and Charles Murray published *The Bell Curve: Intelligence and Class Structure in American Life* (1994). Herrnstein and Murray expressed their beliefs that between 40% and 80% of intelligence is determined by genetics and that it is intelligence levels, not environmental circumstances, poverty, or lack of education, that are at the root of many of our social problems. Critics argued that Herrnstein and Murray not only manipulated and misinterpreted data to support their contention that intelligence levels differ among ethnic groups but also reintroduced outdated and harmful racial stereotypes. Many observers did agree with Herrnstein and Murray's premises that intellect is spread unevenly among individuals and population subgroups, that innate intelligence is distributed through the entire population on a "bell curve," with most people near to the average and fewer at the high and low ends, and even that the distribution varies by race and ethnicity. However, few have been willing to accept the idea that intelligence is entirely genetically encoded, permanently fixed, and unresponsive to environmental influences.

One traditional measure of intelligence is a standardized intelligence quotient (IQ) test, which measures an individual's ability to reason and solve problems. Nearly all studies that focus on the link between intelligence and genetics rely on results obtained from IQ tests, which generally provide an overall score along with measures of verbal ability and performance ability. Although there are several versions of IQ tests available, test takers generally perform comparably on all of them, presumably indicating that they all measure similar aspects of cognitive ability. Critics of IQ tests as measures of intelligence contend that they do not measure all abilities in the complex realm of intelligence. They observe that the tests measure one small segment of the diverse abilities that comprise intelligence, evaluate only analytic abilities, fail to assess creative or practical abilities, and measure only a small sample of the skills that define the domain of intelligent human behavior.

Despite concern about whether IQ tests fully measure intelligence, nearly all scientific study of the contributions of genetics and environment to intelligence has focused on measuring IQ and examining individuals who differ in their familial relationships. For example, if genetics plays the predominant role in IQ, then identical twins should have IQs that compare more closely than the IQs of fraternal twins, and siblings' IQs should correlate more highly than those of cousins. In "Genetics of Childhood Disorders, II: Genetics and Intelligence" (*Journal of the American Academy of Child and Adolescent Psychiatry*, May 1999), Alan S. Kaufman observes that scientists who argue in favor of genetic determination of IQ cite these data to support their assertion that:

- Identical (monozygotic) twins' IQs are more similar than those of fraternal (dizygotic) twins.

- IQs of siblings correlate more highly than IQs of half-siblings, which, in turn, correlate higher than IQs of cousins.

- IQ correlations between a biological parent and child living together are higher than those between an adoptive parent and child living together.

Kaufman asserts that the following results support scientists who believe that environment more strongly determines IQ:

- IQs of fraternal twins correlate more highly than IQs of siblings of different ages despite the same degree of genetic similarity.

- Unrelated siblings reared together, such as biological and adopted children, have IQs that are more similar than biological siblings reared apart.

- Correlations between IQs of an adoptive parent and a child living together are similar to correlations of a biological parent and a child living apart.

- Siblings reared together have IQs that are more similar than siblings reared apart, and the same finding holds for parents and children, when they live together or apart.

The preponderance of evidence from twin, family, and adoption studies supports increasing heritability of intelligence over time, ranging from 20% in infancy to 60% in adulthood, along with environmental factors estimated to contribute about 30%. Kaufman suggests that heredity is important in determining a person's IQ, but environment is also crucial. Based on twin studies, the heritability percentage for IQ is approximately fifty, which is comparable to the heritability value for body weight. Kaufman believes that the genetic contribution to weight and intelligence are comparable. He observes that many overweight people have a genetic predisposition for a large frame and a metabolism that promotes weight gain, whereas naturally thin people have the opposite genetic predisposition. Nonetheless, for most people environmental factors such as diet and exercise have a substantial impact on weight. Similarly, genetics and environment interact to determine IQ, and people with genetic, familial relationships such as parents and siblings frequently share common environments.

In "Virtual Twins: New Findings on Within-Family Environmental Influences on Intelligence" (*Journal of Educational Psychology*, September 2000), psychologist Nancy L. Segal examined virtual twins—genetically unrelated siblings (typically adopted) of the same age who were reared together from early infancy—to assess environmental influences on intelligence. The results of this study reveal that the IQ correlation of virtual twins fell considerably below correlations reported for identical twins, fraternal twins, and full siblings. Segal interprets these results as a demonstration of the modest effects of environment on intellectual development and as supporting a predominantly genetic role in determining intelligence.

Robert Plomin details similar findings from adoption studies in "Genetics of Childhood Disorders, III: Genetics and Intelligence" (*Journal of the American Academy of Child and Adolescent Psychiatry*, June 1999). He observes that adoption studies have a substantial estimated heritability, finding that identical twins reared apart are almost as similar for measures of intelligence as identical twins reared together. Plomin also looked at 245 children adopted in the first month of life from the Colorado Adoption Project to see if there is more of a parent-offspring or an adoptive parent–adopted children correlation of IQ scores. He finds significant correlations between biological mothers and their adopted-away children and almost no parent-offspring correlations for adoptive parents and their adopted children, suggesting that family environment shared by parents and offspring does not contribute as strongly as genetic influences to parent-offspring resemblance for selected measures of intelligence.

Scientists Identify the Gene Involved in Human Brain Development

The genetic underpinnings of human intelligence were further supported by the discovery in 2006 of a key gene—HAR1F—that helped the human brain evolve from its primate ancestors. Kerri Smith et al. report in "Honing in on the Genes for Humanity" (*Nature*, August 17, 2006) that after examining forty-nine areas of genetic code that have changed the most between the human and chimpanzee genomes, they pinpointed an area of the human genome that appears to have evolved seventy times faster than the balance of human genetic code. Smith and her colleagues observed eighteen differences in the HAR1F gene that took place in humans but not in chimps and may have prompted the rapid growth of the brain's cerebral cortex (the part of the brain where thought processes occur). It may also explain why human brains are three times the size of chimp brains.

HOW DO GENES INFLUENCE BEHAVIOR AND ATTITUDES?

The question of interest is no longer whether human social behavior is genetically determined; it is to what extent.

—Edward O. Wilson, *On Human Nature* (1978)

Heredity is what sets the parents of a teenager wondering about each other.

—Laurence J. Peter, author of *The Peter Principle* (1968)

Most people accept the premise that genes at least in part influence personality and behavior. Families have long decried a characteristic "bad temper" or "wild streak" appearing in a new generation, or boasted about inherited musical or artistic talents. It seems intuitively correct to assume that some of our behaviors and attitudes, both the desirable and less desirable ones, are in part genetically mediated. In "The DNA Age: That Wild Streak? Maybe It Runs in the Family" (*New York Times*, June 15, 2006), Amy Harmon asserts that an increasing understanding of genetics has renewed consideration about the degree to which individuals can actually control how they behave.

Studies of families and twins strongly suggest genetic influences on the development and expression of specific behaviors, but there is no conclusive research demonstrating that genes determine behaviors. In "The Interplay of Nature, Nurture, and Developmental Influences: The Challenge ahead for Mental Health" (*Archives of General Psychiatry*, November 2002), psychiatrist Michael Rutter observes that a range of mental health disorders from autism and schizophrenia to attention deficit hyperactivity disorder (ADHD) involve at least indirect genetic effects, with heritability ranging from 20% to 50%. He further asserts that genetically influenced behaviors also bring about gene-environment correlations.

Rutter explains the mechanism of genetic influence on behavior: genes affect proteins, and through the effects of these proteins on the functioning of the brain there are resultant effects on behavior. Rutter views environmental influences as comparable to genetic influences in that they are strong and pervasive but do not determine behaviors, and studies of environmental effects show that there are individual differences in response. Some individuals are severely affected and others experience few repercussions from environmental factors. This has given rise to the idea of varying degrees of resiliency—that people vary in their relative resistance to the harmful effects of psychosocial adversity—as well as the premise that genetics may offer protective effects from certain environmental influences.

Jan Strelau, in "The Contribution of Genetics to Psychology Today and in the Decade to Follow" (*European Psychologist*, December 2001), asserts that the proportion of phenotypic variance that may be attributed to genetic variance shows that personality traits, including temperament as well as specific behaviors and intelligence, have a heritability ranging from 40% to about 60%, but that it is primarily environmental influences that explain individual differences. He states that genetics influence the environment experienced by individuals, which explains how, for example, children growing up in the same family often experience and interpret their environments differently. This also explains why individuals who share the same genes though living apart show some concordance in selecting or creating similar experiences.

Traditional psychological theory holds that attitudes are learned and most strongly influenced by environment. In "The Heritability of Attitudes: A Study of Twins" (*Journal of Personality and Social Psychology*, June 2001), James M. Olson et al. examine whether there is a genetic basis for attitudes by reviewing earlier studies and conducting original research on monozygotic and dizygotic twins. Olson and his collaborators argue that the premise that attitudes are learned is not incompatible with the idea that biological and genetic factors also influence attitudes. They hypothesize that genes probably influence predispositions or natural inclinations, which then shape environmental experiences in ways that increase the likelihood of the individual developing specific traits and attitudes. For example, children who are small for their age might be teased or taunted by other children more than their larger peers. As a result, these children might develop anxieties about social interaction, with consequences for their personalities, such as shyness or low self-esteem discomfort with large groups.

In research supported by the National Institutes of Health, Amy Abrahamson, Laura Baker, and Avshalom Caspi examine genetic influences on attitudes of adolescents and report their findings in "Rebellious Teens? Genetic and Environmental Influences on the Social Attitudes of Adolescents" (*Journal of Personality and Social Psychology*, December 2002). The purpose of their study was to investigate sources of familial influence on adolescent social attitudes in an effort to understand whether and how families exert an influence on the attitudes of adolescents. They wanted to pinpoint the age when genetic influences actually emerge and to determine the extent to which parents and siblings shape teens' views about controversial issues. Abrahamson, Baker, and Caspi explored genetic and environmental influences in social attitudes in 654 adopted and nonadopted children and their biological and adoptive relatives in the Colorado Adoption Project. Conservatism and religious attitudes were measured in the children annually from ages twelve to fifteen and in the parents during the twelve-year-old visit.

The study finds that both conservatism and religious attitudes are strongly influenced by shared-family environmental factors throughout adolescence. Familial resemblance for conservative attitudes arises from both genetic and common environmental factors, and familial influence on religious attitudes is almost entirely in response to shared-family environmental factors. These findings are different from previous findings in twin studies, which suggest that genetic influence on social attitudes do not emerge until adulthood. In contrast, the Colorado Adoption Project study detects significant genetic influence in conservatism as early as age twelve, but finds no evidence of genetic influence on religious attitudes during adolescence. Abrahamson, Baker, and Caspi conclude that genetic factors exert an influence on social attitudes much earlier than previously indicated. For example, significant genetic influences on variations in conservatism are identified as early as age twelve. The study provides further evidence that shared environmental factors contribute significantly to individual differences in social attitudes during adolescence.

Psychologist David Cohen downplays environmental influences by specifically discounting parents' responsibility for mental illness and emotional problems in their children. In *Stranger in the Nest: Do Parents Really Shape Their Child's Personality, Intelligence, or Character?* (1999), Cohen asserts that "good parenting" cannot overcome "bad genes" and that it is impossible to separate genetic background from environmental influence. Making a strong case for genetic influence, Cohen writes, "The truth of the matter is that, if sufficiently strong, inborn potentials can trump parental influence, no matter how positive or negative. Some traits manifest themselves in such unexpected and uncontrollable ways that, for better or for worse, one's child may indeed seem like a perfect stranger."

ENVIRONMENTAL GENOME PROJECT

In many instances it is difficult to understand how genes manage to assert themselves over the myriad of complications imposed by the environment. The Environmental Genome Project (EGP) was launched by the National Institute of Environmental Health Sciences (which is one of the twenty-seven institutes and centers of the National Institutes of Health) in 1998. The EGP aims to improve understanding of human genetic susceptibility to environmental exposures, including gaining an understanding of how individuals differ in their susceptibility to environmental agents and how these susceptibilities change

FIGURE 4.1

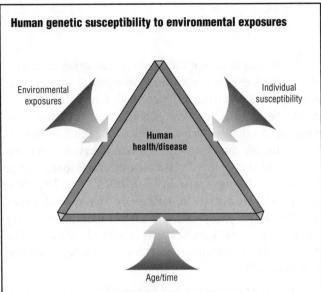

Human genetic susceptibility to environmental exposures

SOURCE: "The Environmental Genome Project (EGP)," in *Environmental Genome Project*, U.S. Department of Health and Human Services, National Institutes of Health, National Institute of Environmental Health Sciences, May 7, 2004, http://www.niehs.nih.gov/envgenom/home.htm (accessed December 8, 2006)

over time. Figure 4.1 shows the interrelationship of human health and disease to environmental exposures, susceptibility, time, and age.

Ongoing EGP research activities focus on biostatistics and bioinformatics; DNA sequencing; ethical, legal, and social implications; population-based epidemiology; and technology development. For example, research underway at the University of Colorado Health Sciences Center is examining controversies surrounding genetic research and health services with Native Americans. This research aims to help formulate guidelines for the conduct of genetic research and the delivery of genetic health services in Native American communities. Other research examines the potential link between environmental exposures and childhood leukemia (cancer of the blood), the health risks of pesticide exposures among farmers, and genetic susceptibility to cancers triggered by environmental exposures such as vinyl chloride induced cancers.

CHAPTER 5
GENETIC DISORDERS

We could wish that ... life-histories were found in every family, showing the health and diseases of its different members. We might thus in time find evidences of pathological connections and morbid liabilities not now suspected.

—William Gull, 1896

It has long been known that heredity affects health. Genetics, the study of single genes and their effects on the body and mind, explains how and why certain traits such as hair color and blood types run in families. Genomics, a discipline that is only about two decades old, is the study of more than single genes; it considers the functions and interactions of all the genes in the genome. In terms of health and disease, genomics has a broader and more promising range than genetics. The science of genomics relies on knowledge of and access to the entire genome and applies to common conditions, such as breast and colorectal cancer, Parkinson's disease, and Alzheimer's disease. It also has a role in infectious diseases once believed to be entirely environmentally caused such as the human immunodeficiency virus (HIV) and tuberculosis. Like most diseases, these frequently occurring disorders result from the interactions of multiple genes and environmental factors. Genetic variations in these disorders may have a protective or a causative role in the expression of diseases.

It is commonly accepted that diseases fall into one of three broad categories: those few that are primarily genetic in origin; those that are largely attributable to environmental causes; and those—most conditions—in which genetics and environmental factors make comparable, though not necessarily equal, contributions. As understanding in genomics advances and scientists identify genes involved in more diseases, the distinctions between these three classes of disorders are diminishing. This chapter considers some of the disorders believed to be predominantly genetic in origin and some that are the result of genes acted on by environmental factors.

There are two types of genes: dominant and recessive. When a dominant gene is passed on to offspring, the feature or trait it determines will appear regardless of the characteristics of the corresponding gene on the chromosome inherited from the other parent. If the gene is recessive, the feature it determines will not show up in the offspring unless both the parents' chromosomes contain the recessive gene for that characteristic. Similarly, among diseases and conditions primarily attributable to a gene or genes, there are autosomal dominant disorders and autosomal recessive disorders.

Another way to characterize genetic disorders is by their pattern of inheritance, as single gene, multifactorial, chromosomal, or mitochondrial. Single-gene disorders (also called Mendelian or monogenic) are caused by mutations in the deoxyribonucleic acid (DNA) sequence of one gene. Because genes code for proteins, when a gene is mutated so that its protein product can no longer carry out its normal function, it may produce a disorder. According to the Human Genome Project Information Web site (December 9, 2003, http://www.ornl.gov/sci/techresources/Human_Genome/medicine/assist.shtml), which is operated by the Department of Energy, there are an estimated 6,000 known single-gene disorders, which occur in about 1 in every 200 births. Examples are cystic fibrosis, sickle-cell anemia, Huntington's disease, and hereditary hemochromatosis (a disorder in which the body absorbs too much iron from food; rather than the excess iron being excreted, it is stored throughout the body, and iron deposits damage the pancreas, liver, skin, and other tissues). Figure 5.1 shows the cystic fibrosis gene and its location on chromosome 7; Figure 5.2 shows the sickle-cell anemia gene found on chromosome 11; and Figure 5.3 shows the hereditary hemochromatosis gene located on chromosome 6. Single-gene disorders are the result of either autosomal dominant, autosomal recessive, or X-linked inheritance.

Multifactorial or polygenic disorders result from a complex combination of environmental factors and mutations in multiple genes. For example, different genes

FIGURE 5.1

FIGURE 5.2

Chromosome 7 and the CFTR gene associated with cystic fibrosis

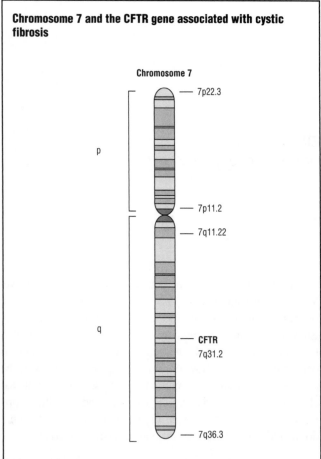

SOURCE: "CFTR: The Gene Associated with Cystic Fibrosis," in "Gene Gateway—Exploring Genes and Genetic Disorders," in *Human Genome Project Information*, U.S. Department of Energy Office of Science, Office of Biological and Environmental Research, Human Genome Project, http://www.ornl.gov/TechResources/Human_Genome/posters/chromosome/cftr.html (accessed October 19, 2006)

Chromosome 11 and the HBB gene associated with sickle-cell anemia

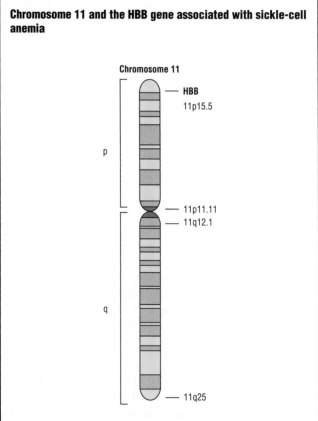

SOURCE: "HBB: The Gene Associated with Sickle-Cell Anemia," in "Gene Gateway—Exploring Genes and Genetic Disorders," in *Human Genome Project Information*, U.S. Department of Energy Office of Science, Office of Biological and Environmental Research, Human Genome Project, http://www.ornl.gov/TechResources/Human_Genome/posters/chromosome/hbb.html (accessed October 19, 2006)

that influence breast cancer susceptibility have been found on seven different chromosomes, rendering it more difficult to analyze than single-gene or chromosomal disorders. Some of the most common chronic diseases are multifactorial in origin. Examples include heart disease, Alzheimer's disease, arthritis, diabetes, and cancer.

Chromosomal disorders are produced by abnormalities in chromosome structure, missing or extra copies of chromosomes, or errors such as translocations (movement of a chromosome section from one chromosome to another). Down syndrome or trisomy 21 is a chromosomal disorder that results when an individual has an extra copy, or a total of three copies, of chromosome 21. Mitochondrial disorders result from mutations in the nonchromosomal DNA of mitochondria, which are organelles involved in cellular respiration. Compared to the three other patterns of inheritance, mitochondrial disorders occur infrequently.

There are significant differences between the nineteenth-century germ theory of disease and the twenty-first-century genomic theory of disease. By the middle of the twentieth century it became possible to improve the quality of life and to save the lives of people with some genetic diseases. Effective treatment included changes in diet to prevent or manage conditions such as phenylketonuria (PKU) and glucose galactose malabsorption (GGM). PKU is an inherited error of metabolism caused by a deficiency in the enzyme phenylalanine hydroxylase. (See Figure 5.4.) It may result in mental retardation, organ damage, and unusual posture. Dietary changes are also used to treat GGM, a rare metabolic disorder caused by lack of the enzyme that converts galactose into glucose. For people with severe cases of GGM, it is vital to avoid lactose (milk sugar), sucrose (table sugar), glucose, and galactose. Other therapeutic measures may involve surgery to correct deformities and avoidance of environmental triggers, as in some types of asthma.

At the dawn of the twenty-first century, the possibility of preventing and changing genetic legacies appears within reach of modern medical science. Genomic medicine predicts the risk of disease in the individual, whether highly probable, as in the case of some of the well-established single-gene disorders, or in terms of an increased suscept-

FIGURE 5.3

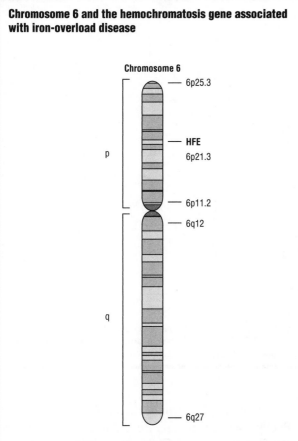

Chromosome 6 and the hemochromatosis gene associated with iron-overload disease

Chromosome 6

6p25.3

HFE
6p21.3

6p11.2

6q12

6q27

p

q

SOURCE: "The Hemochromatosis Gene," in "Gene Gateway—Exploring Genes and Genetic Disorders," in *Human Genome Project Information*, U.S. Department of Energy Office of Science, Office of Biological and Environmental Research, Human Genome Project, http://www.ornl.gov/TechResources/Human_Genome/posters/chromosome/hfe.html (accessed October 19, 2006)

FIGURE 5.4

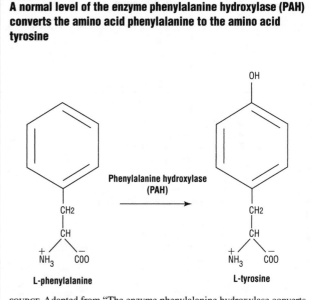

A normal level of the enzyme phenylalanine hydroxylase (PAH) converts the amino acid phenylalanine to the amino acid tyrosine

Phenylalanine hydroxylase (PAH)

L-phenylalanine   L-tyrosine

SOURCE: Adapted from "The enzyme phenylalanine hydroxylase converts the amino acid phenylalanine to tyrosine," in "Nutritional and Metabolic Diseases," in *Genes and Disease*, National Institutes of Health, U.S. National Library of Medicine, National Center for Biotechnology Information, http://www.ncbi.nlm.nih.gov/entrez/query.fcgi?cmd=Search&db=books&doptcmdl=GenBookHL&term=phenylketonuria+AND+gnd%5Bbook%5D+AND+138088%5Buid%5D&rid=gnd.section.234 (accessed December 12, 2006)

ibility likely to be influenced by environmental factors. The promise of genomic medicine is to make preventive medicine more powerful and treatment more specific to the individual, enabling investigation and treatment that are custom-tailored to an individual's genetic susceptibilities, or to the characteristics of the specific disease or disorder.

COMMON GENETICALLY INHERITED DISEASES

Although many diseases, disorders, and conditions are termed *genetic*, classifying a disease as genetic simply means that there is an identified genetic component to either its origin or its expression. Many medical geneticists contend that most diseases cannot be classified as strictly genetic or environmental. The phenotype of genetic diseases can some-times be modified, even to the point of nonexpression, by controlling environmental factors. Similarly, environmental (infectious) diseases may not be expressed because of some genetic predisposition to immunity. Each disease, in each individual, exists along a continuum between a genetic disease and an environmental disease.

A myriad of diseases are believed to have strong genetic contributions, including:

• Heart disease—coronary atherosclerosis, hypertension (high blood pressure), and hyperlipidemia (elevated blood levels of cholesterol and other lipids)

• Diabetes

• Cancer—retinoblastomas, colon, stomach, ovarian, uterine, lung, bladder, breast, skin (melanoma), pancreatic, and prostate

• Neurological disorders—Alzheimer's disease, amyotrophic lateral sclerosis (also known as Lou Gehrig's disease), Gaucher's disease, Huntington's disease, multiple sclerosis, narcolepsy, neurofibromatosis, Parkinson's disease, Tay-Sachs disease, and Tourette's syndrome (a syndrome is a set of symptoms or conditions that taken together suggest the presence of a specific disease or an increased risk of developing the disease)

• Mental illnesses, mental retardation, and behavioral conditions—alcoholism, anxiety disorders, attention deficit hyperactivity disorder, eating disorders, Lesch-Nyhan syndrome, and schizophrenia

• Other genetic disorders—cleft lip and cleft palate, clubfoot, cystic fibrosis, Duchenne muscular dystrophy, glucose galactose malabsorption, hemophilia, Hurler's syndrome,

TABLE 5.1

The fifteen leading causes of death, 2003–04

[Rates of death per 100,000 people; columns may not add up to totals because of rounding]

Rank	Cause of death	Number	Death rate	Age-adjusted death rate		Percent change
				2004	2003	
	All causes	2,398,365	816.7	801.1	832.7	−3.8
1	Diseases of heart	654,092	222.7	217.5	232.3	−6.4
2	Malignant neoplasms	550,270	187.4	184.6	190.1	−2.9
3	Cerebrovascular diseases	150,147	51.1	50.0	53.5	−6.5
4	Chronic lower respiratory diseases	123,884	42.2	41.8	43.3	−3.5
5	Accidents (unintentional injuries)	108,694	37.0	36.6	37.3	−1.9
6	Diabetes mellitus	72,815	24.8	24.4	25.3	−3.6
7	Alzheimer's disease	65,829	22.4	21.7	21.4	1.4
8	Influenza and pneumonia	61,472	20.9	20.4	22.0	−7.3
9	Nephritis, nephrotic syndrome and nephrosis	42,762	14.6	14.3	14.4	−0.7
10	Septicemia	33,464	11.4	11.2	11.6	−3.4
11	Intentional self-harm (suicide)	31,647	10.8	10.7	10.8	−0.9
12	Chronic liver disease and cirrhosis	26,549	9.0	8.8	9.3	−5.4
13	Essential (primary) hypertension and hypertensive renal disease	22,953	7.8	7.6	7.4	2.7
14	Parkinson's disease	18,018	6.1	6.1	6.2	−1.6
15	Pneumonitis due to solids and liquids	16,959	5.8	5.6	5.9	−5.1
	All other causes	418,810	142.6	—	—	

—Category not applicable.

SOURCE: Adapted from "Table B. Deaths and Death Rates for 2004 and Age-Adjusted Death Rates and Percentage Changes in Age-Adjusted Rates from 2003 to 2004 for the 15 Leading Causes of Death in 2004: United States, Final 2003 and Preliminary 2004," in "Deaths: Preliminary Data for 2004," *National Vital Statistics Reports*, vol. 54, no. 19, June 28, 2006, http://www.cdc.gov/nchs/data/nvsr/nvsr54/nvsr54_19.pdf (accessed October 19, 2006)

Marfan's syndrome, phenylketonuria, sickle-cell disease, and thalassemia

- Other medical conditions—including alpha-1-anti-trypsin, arthritis, asthma, baldness, congenital adrenal hyperplasia, migraine headaches, obesity, periodontal disease, porphyria, and some speech disorders

ALZHEIMER'S DISEASE

Alzheimer's disease (AD) is a progressive, degenerative disease that affects the brain and results in severely impaired memory, thinking, and behavior. The National Center for Biotechnology Information (NCBI) reports in *Genes and Disease* (2006, http://www.ncbi.nlm.nih.gov/books/bv.fcgi?call=bv.View..ShowSection&rid=gnd.section .193) that it is the fourth leading cause of death in adults, and the incidence of the disease rises with age. AD affects an estimated four million American adults and is the most common form of dementia, or loss of intellectual function. The National Institute of Mental Health (one of the National Institutes of Health), in "The Numbers Count" (December 26, 2006, https://www .nimh.nih.gov/publicat/numbers.cfm#Alzheimer), estimates that 4.5 million Americans suffer from AD. In 2004 it was the seventh-leading cause of death in the United States. (See Table 5.1.) AD also contributes to many more deaths that are attributed to other causes, such as heart and respiratory failure.

AD has become a disease of particular concern in the United States because the nation's older adult population is growing rapidly. The University of North Texas Center for Public Service reports in "Alzheimer's Disease: Statistics" (August 13, 2001, http://www.cps.unt.edu/ alzheimers/disease_statistics.htm) that approximately 10% of the population over age sixty-five is afflicted with AD. By 2050 the United States will have approximately 86.7 million people over age sixty-five, according to projections released by the U.S. Census Bureau (March 18, 2004, http://www.census.gov/ipc/www/usinterimproj/). The Alzheimer's Association estimates that between 2000 and 2025 the number of cases of Alzheimer's disease in the United States will increase by 44%, with the largest growth in southern and western states that have high populations of retirees (June 7, 2004, http://www.alz.org/documents/ national/FSADState_Growth.pdf). Prevalence (the number of people with a disease at a given time) is partially determined by the length of time people with AD survive. Even though the average survival is eight years after diagnosis, some AD patients have lived longer than twenty years with the disease. Therefore, improvements in AD care, as well as increased length of life of the older adult population in general, will increase the numbers of AD patients.

Genetic Causes of AD

AD is not a normal consequence of growing older, and scientists are continuing to seek its cause. Researchers find some promising genetic clues to the disease. Table 5.2 shows the different patterns of inheritance, ages of onset (when symptoms begin), genes, chromosomes, and proteins linked to the development of AD. Mutations in four genes,

TABLE 5.2

Genes for Alzheimer's disease

Age at onset	Inheritance	Chromosome	Gene	Protein	% AD
Early onset	AD	14	PS1	Presenilin 1	<2
Early onset	AD	21	APP	Amyloid Precursor protein	<20 families*
Early onset	AD	1	PS2	Presenilin 2	3 families*
Early onset	AD	?	?	?	?
Late onset	Familial/sporadic	19	APOE	Apolipoprotein E	~50
Late onset	Familial	12p11-q13	?	?	?
Late onset	Familial	9p22.1	?	?	?
Late onset	Familial	10q24	?	?	?
Late onset	?	?	?	?	?

Age of onset: early onset: <60 years, late onset: >60 years; inheritance: AD: autosomal dominant, familial: disease in at least one first-degree relative, sporadic: disease in no other family member; chromosome: number, arm, and region; gene: designation of identified gene; protein: name of protein coded for by the gene; % AD: percent of AD caused by or *number of families identified with AD for each gene.

SOURCE: Richard Robinson, ed., "Genes for Alzheimer's Disease," in *Genetics*, Vol. 1, *A–D*, Macmillan Reference USA, 2002

situated on chromosomes 1, 14, 19, and 21, are thought to be involved in the disease, and the best described are PS1 (or AD3) on chromosome 14 and PS2 (or AD4) on chromosome 1. (See Figure 5.5 and Figure 5.6.)

The formation of lesions made of fragmented brain cells surrounded by amyloid-family proteins is characteristic of the disease. Interestingly, these lesions and their associated proteins are closely related to similar structures found in Down syndrome. Tangles of filaments largely made up of a protein associated with the cytoskeleton have also been observed in samples taken from AD brain tissue.

The first genetic breakthrough was reported in the February 1991 issue of the British journal *Nature*. Investigators reported that they had discovered that a mutation in a single gene could cause this progressive neurological illness. Scientists found the defect in the gene that directs cells to produce a substance called amyloid protein. Researchers at the Massachusetts Institute of Technology found that low levels of the brain chemical acetylcholine contribute to the formation of hard deposits of amyloid protein that accumulate in the brain tissue of AD patients. In unaffected people the protein fragments are broken down and excreted by the body. Amyloid protein is found in cells throughout the body. Researchers do not know why it becomes a deadly substance in the brain cells of some people and not others.

In 1995 three more genes linked to AD were identified. One gene appears to be related to the most devastating form of AD, which can strike people in their thirties. When defective, the gene may prevent brain cells from correctly processing a substance called beta amyloid precursor protein. The second gene is linked to another early-onset form of AD that strikes before age sixty-five. This gene also appears to be involved in producing beta amyloid. Researchers believe that the discovery of these two genes will allow them to narrow their search for the proteins responsible for early-onset AD and give them clues to the causes of AD in older people. (See Table 5.2.)

According to the National Center for Biotechnology Information (2007, http://www.ncbi.nlm.nih.gov/entrez/dispomim.cgi?id=107741), the third gene, known as apolipoprotein E (apoE), was actually reported as associated with AD in 1993, but its role in the body was not known at that time. Researchers have since found that the gene plays several roles. It regulates lipid metabolism within the organs and helps to redistribute cholesterol. In the brain apoE participates in repairing nerve tissue that has been injured. There are three forms (alleles) of the gene: apoE-2, apoE-3, and apoE-4. Until recently, people with two copies of apoE-4, one from each parent, were thought to have a greatly increased risk of developing AD before age seventy. Between one-half and one-third of all AD patients have at least one apoE-4 gene, whereas only 15.5% of the general population have an apoE-4 gene. In 1998, however, researchers discovered that the apoE-4 gene seems to affect when a person may develop AD, not whether the person will develop the disease.

Another newly discovered gene, A2M-2, appears to affect whether a person will develop AD. The article "Late-Onset Alzheimer's Gene Suggests Interplay" (*Neuroscience*, August 14, 1998) indicates that nearly one-third (30%) of Americans may carry A2M-2, a genetic variant that more than triples their risk of developing late-onset AD compared to siblings with the normal version of the A2M gene. The discovery of A2M-2 opens up the possibility of developing a drug that mimics the A2M gene's normal function. This can protect susceptible people against brain damage or perhaps even reverse it.

Testing for AD

A complete physical, psychiatric, and neurological evaluation can usually produce a diagnosis of AD that is about 90% accurate. For many years the only sure way to diagnose the disease was to examine brain tissue under a microscope, which was not possible while the AD victim was still alive. An autopsy of someone who has died of

FIGURE 5.5

Chromosome 1 and the AD4 gene associated with Alzheimer's disease

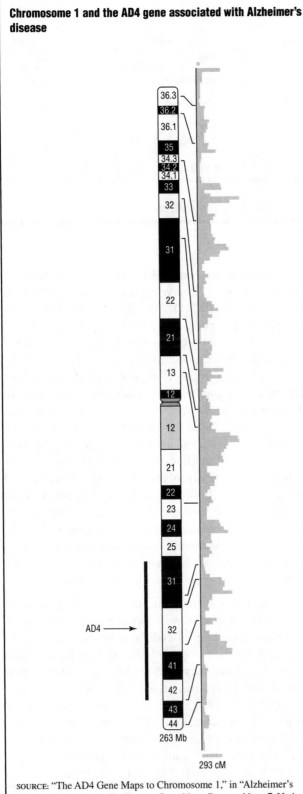

SOURCE: "The AD4 Gene Maps to Chromosome 1," in "Alzheimer's Disease," in *Science: The Human Gene Map, Genome Maps 7*, National Institutes of Health, U.S. National Library of Medicine, National Center for Biotechnology Information, http://www.ncbi.nlm.nih.gov/SCIENCE96/gene.cgi?AD4 (accessed October 19, 2006)

FIGURE 5.6

Chromosome 14 and the AD3 gene associated with Alzheimer's disease

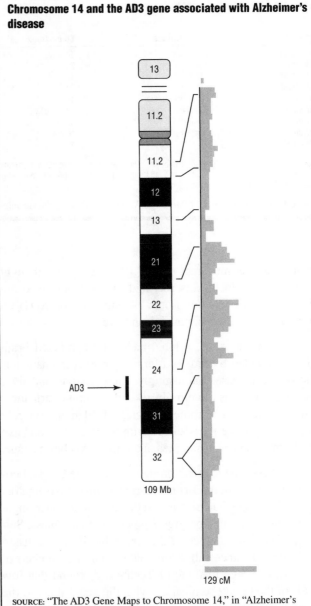

SOURCE: "The AD3 Gene Maps to Chromosome 14," in "Alzheimer's Disease," in *Science: The Human Gene Map, Genome Maps 7*, National Institutes of Health, U.S. National Library of Medicine, National Center for Biotechnology Information, http://www.ncbi.nlm.nih.gov/SCIENCE96/gene.cgi?AD3 (accessed October 19, 2006)

AD reveals a characteristic pattern that is the hallmark of the disease: tangles of fibers (neurofibrillary tangles) and clusters of degenerated nerve endings (neuritic plaques) in areas of the brain that are crucial for memory and intellect. Also, the cortex of the brain is shrunken.

Diagnostic tests for AD have included analysis of blood and spinal fluid as well as the use of magnetic resonance imaging (MRI) to measure the volume of brain tissue in areas of the brain used for memory, organizational ability, and planning to accurately identify people with AD and predict who will develop AD in the future.

In "Nanoparticle-Based Detection in Cerebral Spinal Fluid of a Soluble Pathogenic Biomarker for Alzheimer's Disease" (*Proceedings of the National Academy of Science*, February 15, 2005), Dimitra G. Georganopoulou et al. report development of another diagnostic test that detects small amounts of protein in spinal fluid. The test is called a bio-barcode assay and is as much as a million times more sensitive than other tests. First used to identify a marker for prostate cancer, the test is used to detect a protein in the brain called amyloid beta-derived diffusible ligand (ADDL). ADDLs are small soluble proteins that may be indicative of AD. To detect them, Georganopoulou and her colleagues use nanoscale particles that have antibodies specific to ADDL.

Physicians and neuroscientists have been eager for a simple and accurate test that can distinguish people with AD from those with cognitive problems or dementias arising from other causes. An accurate test would allow the detection of AD early enough for the use of experimental medications to slow the progression of the disease, as well as identify those at risk of developing AD. However, the availability of tests raises ethical and practical questions: Do patients really want to know their risks of developing AD? Will health insurers use genetic or other diagnostic test results to deny insurance coverage?

Treatments for AD

There is still no cure or prevention for AD, and treatment focuses on managing symptoms. Medication can lessen some of the symptoms, such as agitation, anxiety, unpredictable behavior, and depression. Physical exercise and good nutrition are important, as is a calm and highly structured environment. The object is to help the AD patient maintain as much comfort, normalcy, and dignity for as long as possible.

By 2005 five prescription drugs were available to treat people who suffer from AD. Four of these—galantamine, rivastigmine, donepezil, and tacrine—are cholinesterase inhibitors and are prescribed for the treatment of mild to moderate AD. These drugs produce some delay in the deterioration of memory and other cognitive skills in some patients. They offer mild benefits at best and may lose their effectiveness over time, but currently they are the only alternatives available to treat mild to moderate AD.

The fifth approved medication, memantine, is prescribed for the treatment of moderate to severe AD. It acts to delay progression of some symptoms of moderate to severe AD and may allow patients to maintain certain daily functions a little longer. It is thought to work by regulating glutamate, a chemical in the brain that, in excessive amounts, may lead to brain cell death.

Other researchers are examining the roles of the hormones estrogen and progesterone on memory and cognitive function. Because AD involves inflammatory processes in the brain, scientists are also studying the use of anti-inflammatory agents such as ibuprofen and prednisone to reduce the risk of developing AD. Researchers are also investigating the relationship between the various gene sites, particularly the mutation on chromosome 21, and environmental influences that may increase susceptibility to AD. Furthermore, researchers are trying to determine whether antioxidants, such as vitamin E, can prevent people with mild memory impairment from progressing to AD.

CANCER

Cancer is a large group of diseases characterized by uncontrolled cell division and the growth and spread of abnormal cells. These cells may grow into masses of tissue called tumors. Tumors composed of cells that are not cancerous are called benign tumors. Tumors consisting of cancer cells are called malignant tumors. The dangerous aspect of cancer is that cancer cells invade and destroy normal tissue.

The mechanisms of action that disrupt the cell cycle are impairment of a DNA repair pathway, transformation of a normal gene into an oncogene (a hyperactive gene that stimulates cell growth), and the malfunction of a tumor-suppressor gene (a gene that inhibits cell division). Figure 5.7 shows the multiple systems that interact to control the cell cycle.

The spread of cancer cells occurs either by local growth of the tumor or by some of the cells becoming detached and traveling through the blood and lymphatic system to seed additional tumors in other parts of the body. Metastasis (the spread of cancer cells) may be confined to a local region of the body, but if left untreated (and often despite treatment), the cancer cells can spread throughout the entire body, eventually causing death. It is perhaps the rapid, invasive, and destructive nature of cancer that makes it, arguably, the most feared of all diseases, even though it is second to heart disease as the leading cause of death in the United States. (Table 5.1 uses the term *malignant neoplasms* to describe cancer.)

Cancer can be caused by both external environmental influences (chemicals, radiation, and viruses) and internal factors (hormones, immune conditions, and inherited mutations). These factors may act together or in sequence to begin or promote cancer. There is consensus in the scientific community that several cancer-promoting influences accrue and interact before an individual will develop a malignant growth. With only a few exceptions, no single factor or risk alone is sufficient to cause cancer. As with other disorders that arise in response to multiple factors, susceptibility to certain cancers is often attributed to a mutated gene. Figure 5.8 shows how the inheritance of a mutated gene increases susceptibility for retinoblastoma (cancer of the eye that affects approximately 300 children in the United States each year).

FIGURE 5.7

The cell cycle and cancer

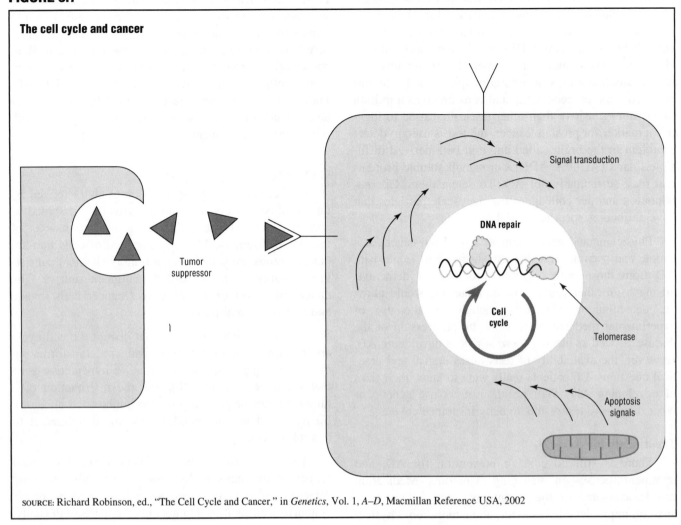

SOURCE: Richard Robinson, ed., "The Cell Cycle and Cancer," in *Genetics*, Vol. 1, *A–D*, Macmillan Reference USA, 2002

Genetic Research

Scientists and physicians have known for some time that predisposition to some forms of breast cancer are inherited and have been searching for the gene or genes responsible so that they can test patients and provide more careful monitoring for those at risk. In 1994 doctors identified the BRCA1 gene, and in late 1995 they also isolated the BRCA2 gene. Since then it has been found that variations of the ATM, BRCA1, BRCA2, CHEK2, and RAD51 genes increase the risk of developing breast cancer, and the AR, DIRAS3, and ERBB2 (also called Her-2/neu) genes are associated with breast cancer.

Viviana Rivera-Varas, in "Breast Cancer Genes and Inheritance" (1998, http://www.cc.ndsu.nodak.edu/instruct/mcclean/plsc431/students98/rivera.htm), notes that if a woman with a family history of breast cancer inherits a defective form of either BRCA1 or BRCA2, she has an estimated 80% to 90% chance of developing breast cancer. Researchers also think that the two genes are linked to ovarian, prostate, and colon cancer and that BRCA2 likely plays some role in breast cancer in men. Scientists suspect that the two genes may also

participate in some way in the development of breast cancer in women with no family history of the disease. Sung-Won Kim et al. state in "Prevalence of BRCA2 Mutations in a Hospital Based Series of Unselected Breast Cancer Cases" (*Journal of Medical Genetics*, 2005) that only about 10% of all cases of breast cancer are attributable to the susceptibility genes BRCA1 and BRCA2. Figure 5.9 shows the location of the BRCA1 gene on the long arm of chromosome 17 at position 21. Figure 5.10 shows the location of the BRCA2 gene on the long arm of chromosome 13 at position 12.3.

According to Els M. Berns et al. in "Oncogene Amplification and Prognosis in Breast Cancer: Relationship with Systemic Treatment" (*Gene*, June 14, 1995), another form of breast cancer, due to multiple copies of a gene called ERBB2, causes an estimated 25% of the approximately 213,000 new cases of the disease in the United States each year. ERBB2 often triggers an aggressive form of cancer that can cause death more quickly than other breast cancers, often within ten to eighteen months after the cancer spreads. The ERBB2 gene produces a protein on the surface of cells that serves as a receiving point for growth-stimulating hormones. Figure 5.11 shows

FIGURE 5.8

Inheritance of a mutated retinoblastoma gene

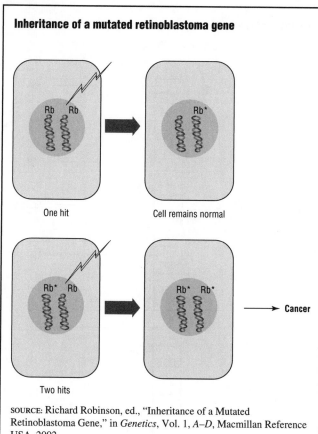

SOURCE: Richard Robinson, ed., "Inheritance of a Mutated Retinoblastoma Gene," in *Genetics*, Vol. 1, *A–D*, Macmillan Reference USA, 2002

FIGURE 5.9

Location of the breast cancer 1, early onset gene (BRCA1)

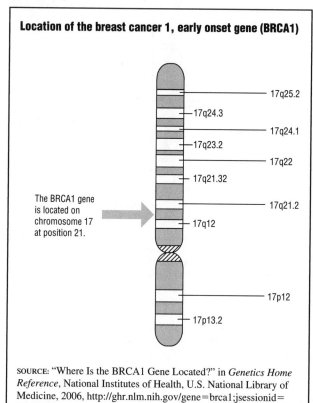

SOURCE: "Where Is the BRCA1 Gene Located?" in *Genetics Home Reference*, National Institutes of Health, U.S. National Library of Medicine, 2006, http://ghr.nlm.nih.gov/gene=brca1;jsessionid=B9C5F23B7BFED46324978C8EC00D7DF1#location (accessed October 19, 2006)

FIGURE 5.10

Location of the breast cancer 2, early onset gene (BRCA2)

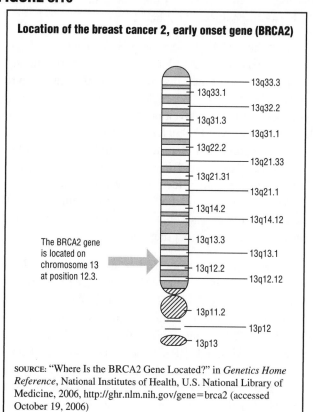

SOURCE: "Where Is the BRCA2 Gene Located?" in *Genetics Home Reference*, National Institutes of Health, U.S. National Library of Medicine, 2006, http://ghr.nlm.nih.gov/gene=brca2 (accessed October 19, 2006)

FIGURE 5.11

Location of the tumor-producing gene ERBB2

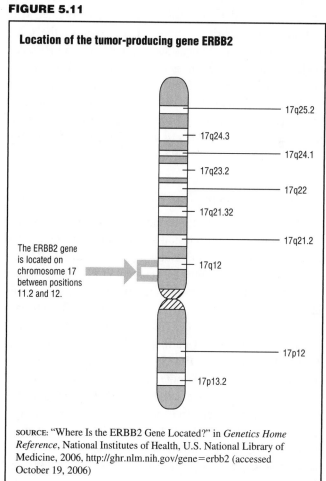

SOURCE: "Where Is the ERBB2 Gene Located?" in *Genetics Home Reference*, National Institutes of Health, U.S. National Library of Medicine, 2006, http://ghr.nlm.nih.gov/gene=erbb2 (accessed October 19, 2006)

the location of ERBB2 on the long arm of chromosome 17 between positions 11.2 and 12.

In "The BARD1 Cys557Ser Variant and Breast Cancer Risk in Iceland" (*PLOS Medicine*, July 2006), Simon N. Stacey et al. report their discovery of another mutation that when present with BRCA1 or BRCA2 significantly increases the risk of developing breast cancer. Stacey and his coauthors studied 1,090 Icelandic women who had breast cancer and compared them to 703 similar women without the disease. They found a specific mutation, BARD1, in 2.8% of women with cancer but only 1.6% of those without cancer. They also found that women with both the BARD1 and BRCA2 mutations were twice as likely to develop breast cancer as women without these mutations.

Another study (Sheila Seal et al., "Truncating Mutations in the Fanconi Anemia J Gene, BRIP1, Are Low Penetrance Breast Cancer Susceptibility Alleles," *Nature Genetics*, November 2006), identifies a new genetic mutation, BRIP1, that doubles the risk of breast cancer in carriers. Although BRIP1 increases a woman's breast cancer risk twofold, other gene mutations raise it much more. Seal and the other researchers find that mutations in BRCA1, BRCA2, and another gene, TP53, increase the carrier's risk of breast cancer by ten- to twentyfold by age sixty. Mutations in other genes such as CHEK2, ATM, and the newly identified BRIP1 gene are associated with a more moderate risk increase. Like some of the other known breast cancer genes such as BRCA1 and BRCA2, BRIP1 is a DNA-repair gene, so women with a faulty version of this gene cannot repair damaged DNA correctly. Individuals with faulty DNA-repair genes have an increased risk of cancer because their healthy cells are more likely to accumulate genetic damage that can trigger the cell to replicate uncontrollably.

CYSTIC FIBROSIS

Cystic fibrosis (CF) is the most common inherited fatal disease of children and young adults in the United States. According to the National Library of Medicine's Genetics Home Reference (January 5, 2007, http://ghr.nlm.nih.gov/condition%3Dcysticfibrosis), CF occurs in about 1 out of 3,200 whites, 1 out of 15,000 African-Americans, and 1 out of 31,000 Asian-Americans. The National Institutes of Health's National Human Genome Research Institute (NHGRI; March 2006, http://www.genome.gov/10001213) notes that approximately 30,000 young people had the disease in 2006; their median life span was thirty years (meaning half would live longer and half would not live this long). An estimated 10 million Americans, almost all of whom are white, are symptomless carriers of the CF gene. Like sickle-cell disease, it is a recessive genetic disorder—to inherit this

FIGURE 5.12

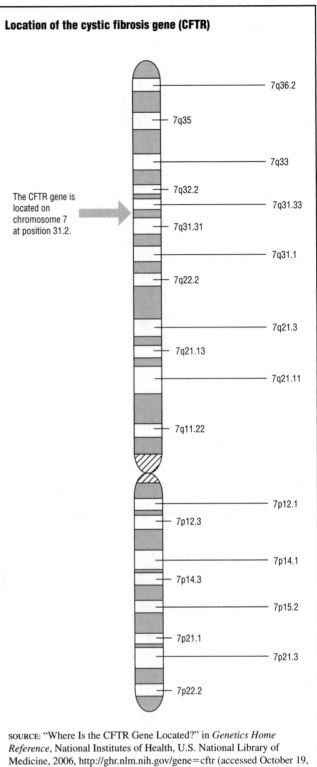

Location of the cystic fibrosis gene (CFTR)

The CFTR gene is located on chromosome 7 at position 31.2.

7q36.2
7q35
7q33
7q32.2
7q31.33
7q31.31
7q31.1
7q22.2
7q21.3
7q21.13
7q21.11
7q11.22
7p12.1
7p12.3
7p14.1
7p14.3
7p15.2
7p21.1
7p21.3
7p22.2

SOURCE: "Where Is the CFTR Gene Located?" in *Genetics Home Reference*, National Institutes of Health, U.S. National Library of Medicine, 2006, http://ghr.nlm.nih.gov/gene=cftr (accessed October 19, 2006)

disease, a child must receive the CF gene from both parents.

In 1989 the CF gene was identified, and in 1991 it was cloned and sequenced. It is located on the long arm of chromosome 7 at position 31.2. (See Figure 5.12.) The gene was called cystic fibrosis transmembrane conductance

FIGURE 5.13

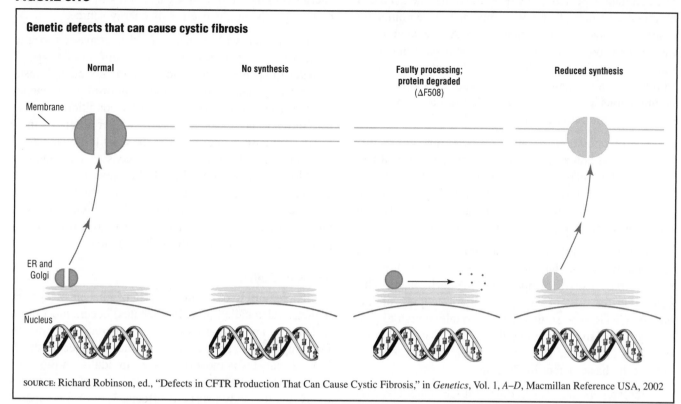

Genetic defects that can cause cystic fibrosis

SOURCE: Richard Robinson, ed., "Defects in CFTR Production That Can Cause Cystic Fibrosis," in *Genetics*, Vol. 1, *A–D*, Macmillan Reference USA, 2002

regulator (CFTR) because it was discovered to encode a membrane protein that controls the transit of chloride ions across the plasma membrane of cells. Nearly 1,000 mutations of the large gene—250,000 nucleotides—have been identified. Though most are extremely rare, several account for more than two-thirds of all mutations. Figure 5.13 shows three types of defects in CFTR that can cause CF. The most frequently occurring mutation causes faulty processing of the protein such that the protein is degraded before it reaches the cell membrane.

The mutated versions of the gene found in people with CF were seen to cause relatively modest impairment of chloride transport in cells. However, this seemingly minor defect can result in a multisystem disease that affects organs and tissues throughout the body, provoking abnormal, thick secretions from glands and epithelial cells. Ultimately, these secretions fill the lungs and cause affected children to die of respiratory failure.

The progression from the defective gene and protein it encodes to life-threatening illness follows this complex path:

1. The defect in chloride passage across the cell membrane indirectly produces an accumulation of thick mucus secretions in the lungs.

2. The bacteria *Pseudomonas aeruginosa* grows in the mucus.

3. In a campaign to combat the bacterial invasion, the body's immune system is activated but is unable to access the bacteria because the thick mucus protects it.

4. The immune reaction persists and becomes chronic, resulting in inflammation that harms the lung.

5. Ultimately, it is the affected individual's own immune response, rather than the defective CF gene, protein, or the bacterial infection, that produces the often fatal damage to the lungs.

A simple sweat test is currently the standard diagnostic test for CF. The test measures the amount of salt in the sweat; abnormally high levels are the hallmark of cystic fibrosis.

CF Gene-Screening Falters

In August 1989 researchers isolated the specific gene that causes CF. The mutation of this gene accounts for about 70% of the cases of the disease. In 1990 scientists successfully corrected the biochemical defect by inserting a healthy gene into diseased cells grown in the laboratory, a major step toward developing new therapies for the disease. In 1992 they injected healthy genes into laboratory rats with a deactivated common cold virus as the delivery agent. The rats began to manufacture the missing protein, which regulates the chloride and sodium in the tissues, preventing the deadly buildup of mucus. Scientists were hopeful that within a few years CF would be eliminated as a fatal disease, giving many children the chance for healthy, normal lives.

In 1993, however, optimism faded when the medical community discovered that the CF gene was more complicated than expected. Scientists found that the gene can be mutated at more than 950 points, and more points are being recognized at an alarming rate. At the same time, they discovered that many people who have inherited mutated genes from both parents do not have cystic fibrosis. With so many possible mutations, the potential combinations in a person who inherits one gene from each parent are immeasurable.

The combinations of different mutations create different effects. Some may result in crippling and fatal CF, whereas others may cause less serious disorders, such as infertility, asthma, or chronic bronchitis. To further complicate the picture, other genes can alter the way different mutations of the CF gene affect the body.

The Cystic Fibrosis Foundation supports clinical research studies in human gene therapy. One form of gene therapy uses a compacted DNA technology. This gene transfer system compacts single copies of the healthy CFTR gene so that they are small enough to pass through a cell membrane into the nucleus. The goal is for the DNA to produce the CFTR protein that is needed to correct the basic defect in CF cells.

Researchers are also finding that CF mutations may be much more common than previously thought. For example, Stephanie A. Cohen and Lyn Hammond, in "Abstracts from the Nineteenth Annual Education Conference of the National Society of Genetic Counselors (Savannah, Georgia, November 2000)" (*Journal of Genetic Counseling*, December 2000), cite a study of 5,000 healthy women receiving prenatal care at Kaiser Permanente in northern California who were tested for the CF gene, which is thought to be present in less than 1% of the population. Of those screened, 11% had the mutation. This finding may indicate that many more common diseases, such as asthma, may be caused by mutations of the CF gene. Other scientists speculate that the frequency of CF carriers among people of European descent may have, at some point in time, conferred immunity to some other disorder, in much the same way that the sickle-cell carriers were protected from malaria.

DIABETES

Diabetes is a disease that affects the body's use of food, causing blood glucose (sugar levels in the blood) to become too high. Normally, the body converts sugars, fats, starches, and proteins into a form of sugar called glucose. The blood then carries glucose to all the cells throughout the body. In the cells, with the help of the hormone insulin, which facilitates the entry of glucose into the cells, the glucose is either converted into energy for use immediately or stored for the future. Beta cells of the pancreas, a small organ located behind the stomach, manufacture the insulin. The process of turning food into energy via glucose (blood sugar) is important because the body depends on glucose for its energy source.

In a person with diabetes, food is converted to glucose, but there is a problem with insulin. In one type of diabetes the pancreas does not manufacture enough insulin, and in another type the body has insulin but cannot use the insulin effectively (this latter condition is called insulin resistance). When insulin is either absent or ineffective, glucose cannot get into the cells to be converted into energy. Instead, the unused glucose accumulates in the blood. If a person's blood-glucose level rises high enough, the excess glucose is excreted from the body via urine, causing frequent urination. This, in turn, leads to an increased feeling of thirst as the body tries to compensate for the fluid lost through urination.

Types of Diabetes

There are two distinct types of diabetes. Type 1 diabetes (also called juvenile diabetes) occurs most often in children and young adults. The pancreas stops manufacturing insulin, so the hormone must be injected daily. Type 2 diabetes is most often seen in adults. In this type the pancreas produces insulin, but it is not used effectively and the body resists its effects. Table 5.3 compares the phenotype (presentation) and genotype of Type 1 and Type 2 diabetes.

According to the Centers for Disease Control and Prevention (CDC; October 26, 2005, http://www.cdc.gov/od/oc/media/pressrel/fs051026.htm), in 2005 approximately 21 million Americans (7% of the population) had diabetes. Of these, approximately one-third had not been diagnosed. In 2004 diabetes was the sixth leading cause of death and in 2005, 1.5 million new cases of diabetes were diagnosed. (See Table 5.1 and Figure 5.14.) The individuals most at risk for Type 2 diabetes are usually overweight, over forty years old, and have a family history of diabetes. The CDC (January 31, 2005, http://www.cdc.gov/diabetes/pubs/general.htm) reports that Type 2 patients represent 90% to 95% of diabetes patients, while Type 1 accounts for 5% to 10% of diabetes cases.

Causes of Diabetes

The causes of both Type 1 and Type 2 diabetes are unknown, but a family history of the disease increases the risk for both types, leading researchers to believe there is a genetic component. Some scientists believe that a flaw in the body's immune system may be a factor in Type 1 diabetes. Poor cardiovascular fitness is another risk factor for developing diabetes.

Mutations in several genes probably contribute to the origin and onset of Type 1 diabetes. For example, an insulin-dependent diabetes mellitus (IDDM1) site on chromosome 6 may harbor at least one susceptibility gene for Type 1 diabetes. The role of this mutation in increasing

TABLE 5.3

Comparison of type 1 and type 2 diabetes

	Type 1 diabetes	Type 2 diabetes
Phenotype (observable characteristics)	Onset primarily in childhood and adolescence Often thin or normal weight Prone to ketoacidosis Insulin administration required for survival Pancreas is damaged by an autoimmune attack Absolute insulin deficiency Treatment: insulin injections	Onset predominantly after 40 years of age* Often obese No ketoacidosis Insulin administration not required for survival Pancreas is not damaged by an autoimmune attack Relative insulin deficiency and/or insulin resistance Treatment: (1) healthy diet and increased exercise; (2) hypoglycemic tablets; (3) insulin injections
Genotype (genetic makeup)	Increased prevalence in relatives Identical twin studies: <50% concordance HLA association: yes	Increased prevalence in relatives Identical twin studies: usually above 70% concordance HLA association: no

*Type 2 diabetes is increasingly diagnosed in younger patients.
Note: HLA is human leukocyte antigen.

SOURCE: Adapted from Laura Dean and Johanna McEntyre, "Table 1. Comparison of Type 1 and Type 2 Diabetes," in *The Genetic Landscape of Diabetes*, U.S. Department of Health and Human Services, National Institutes of Health, National Institute of Diabetes and Digestive and Kidney Diseases, 2004, http://www.ncbi.nlm.nih.gov/books/bv.fcgi?rid=diabetes.table.580 (accessed October 19, 2006)

FIGURE 5.14

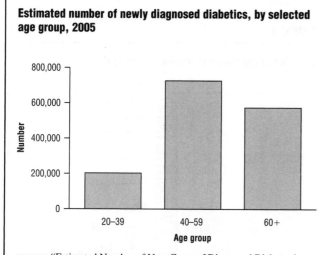

Estimated number of newly diagnosed diabetics, by selected age group, 2005

SOURCE: "Estimated Number of New Cases of Diagnosed Diabetes in People Aged 20 Years or Older by Age Group—United States, 2005," in *National Diabetes Fact Sheet: United States, 2005* U.S. Department of Health and Human Services, Centers for Disease Control and Prevention, 2005, http://www.cdc.gov/diabetes/pubs/pdf/ndfs_2005.pdf (accessed October 19, 2006)

susceptibility is not yet known; however, because chromosome 6 also contains genes for antigens (the molecules that normally tell the immune system not to attack itself), there may be some interaction between immunity and diabetes. In Type 1 diabetes the body's immune system mounts an immunological assault on its own insulin and the pancreatic cells that manufacture it. Some ten sites in the human genome, including a gene at the locus IDDM2 on chromosome 11 and the gene for glucokinase, an enzyme that is crucial for glucose metabolism, on chromosome 7, appear to increase susceptibility to Type 1 diabetes.

In Type 2 diabetes heredity may be a factor, but because the pancreas continues to produce insulin, the disease is considered a problem of insulin resistance, in which the body is not using the hormone efficiently. In people prone to Type 2 diabetes, being overweight can set off the disease because excess fat prevents insulin from working correctly. Maintaining a healthy weight and keeping physically fit can usually prevent noninsulin-dependent diabetes. To date, insulin-dependent diabetes (Type 1) cannot be prevented.

Complications of Diabetes

Because diabetes deprives body cells of the glucose needed to function properly, complications can develop that threaten the lives of diabetics. Complications of diabetes include higher risk and rates of heart disease; circulatory problems, especially in the legs, often severe enough to require surgery or even amputation; diabetic retinopathy, a condition that can cause blindness; kidney disease that may require dialysis; dental problems; impaired healing and increased risk of infection; and problems of pregnancy. People who pay close attention to the roles of diet, exercise, weight management, and pharmacological control (proper use of insulin and other medication) to manage their disease suffer the fewest complications.

HUNTINGTON'S DISEASE

Huntington's disease (HD) is an inherited, progressive brain disorder. It causes the degeneration of cells in the basal ganglia, a pair of nerve clusters deep in the brain that affect both the body and the mind. HD is caused by a single dominant gene that affects men and women of all races and ethnic groups. Figure 5.15 shows the inheritance pattern of HD.

FIGURE 5.15

The inheritance pattern of Huntington's disease (HD)

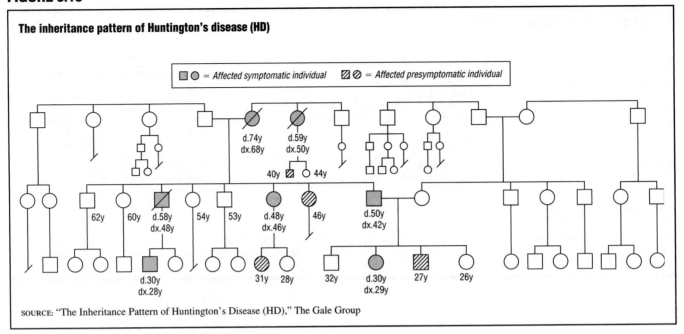

SOURCE: "The Inheritance Pattern of Huntington's Disease (HD)," The Gale Group

Gene Responsible for HD Found

The gene mutation that produces HD was mapped to chromosome 4 in 1983 and cloned in 1993. (See Figure 5.16.) In the HD gene the mutation involves a triplet of nucleotides, cytosine (C), adenine (A), and guanine (G), known as CAG. The mutation is an expansion of a nucleotide triplet repeat in the DNA that codes for the protein huntingtin. In unaffected people the gene has thirty or fewer of these triplets, but HD patients have forty or more. These increased multiples either destroy the gene's ability to make the necessary protein or cause it to produce a misshapen and malfunctioning protein. Either way, the defect results in the death of brain cells.

The number of repeated triplets is inversely related to the age when the individual first experiences symptoms—the more repeated triplets, the younger the age of onset of the disease. Like myotonic dystrophy, in which the symptoms of the disease often increase in severity from one generation to the next, the unstable triplet repeat sequence can lengthen from one generation to the next, with a resultant decrease in the age when symptoms first appear.

HD does not usually strike until mid-adulthood, between the ages thirty and fifty, although there is a juvenile form that can affect children and adolescents. Early symptoms, such as forgetfulness, a lack of muscle coordination, or a loss of balance, are often ignored, delaying the diagnosis. The disease gradually takes its toll over a ten- to twenty-five-year period.

Within a few years characteristic involuntary movement (chorea) of the body, limbs, and facial muscles appears. As HD progresses, speech becomes slurred and swallowing becomes difficult. The patients' cognitive abilities decline,

FIGURE 5.16

Location of the Huntington's disease gene (HD)

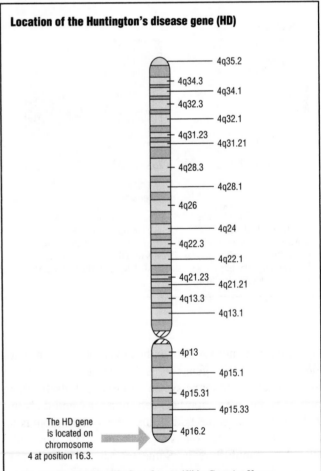

The HD gene is located on chromosome 4 at position 16.3.

SOURCE: "Where Is the HD Gene Located?" in *Genetics Home Reference*, National Institutes of Health, U.S. National Library of Medicine, 2006, http://ghr.nlm.nih.gov/gene=hd (accessed October 19, 2006)

and there are distinct personality changes—depression and withdrawal, sometimes countered with euphoria. Eventually, nearly all patients must be institutionalized, and they usually die as a result of choking or infections.

Prevalence of HD

HD, once considered rare, is now recognized as one of the more common hereditary diseases. According to the Genetics Home Reference (January 5, 2007, http://ghr. nlm.nih.gov/ condition=huntingtondisease), HD affects 3 to 7 per 100,000 people of European ancestry. HD appears to be less common in other populations, including people of Japanese, Chinese, and African descent. The NHGRI (October 2006, http://www.genome.gov/10001215) reports that in the United States about 30,000 people have HD, an additional 35,000 people exhibit some symptoms, and 75,000 people carry the abnormal gene that will cause them to develop the disease.

Prediction Test

In 1983 researchers identified a DNA marker that made it possible to offer a test to determine whether an individual has inherited the HD gene before symptoms appear. In some cases it is even possible to make a prenatal diagnosis on an unborn child. Many people, however, prefer not to know whether or not they carry the defective gene. Currently, researchers are trying to determine if the exact number of excess triplets indicates when in life a person will be affected by the disease. Some scientists fear that the ability to tell people that they are going to develop an incurable disease and pinpoint when they will develop it will make genetic testing, already a difficult decision, even more complicated.

MUSCULAR DYSTROPHY

Muscular dystrophy (MD) is a term that applies to a group of more than thirty types of hereditary muscle-destroying disorders. More than a million Americans are affected by one of the forms of MD. Each variant of the disease is caused by defects in the genes that play important roles in the growth and development of muscles. Duchenne muscular dystrophy (DMD) is one of the most frequently occurring types of MD and is characterized by rapid progression of muscle degeneration that occurs early in life. In all forms of MD the proteins produced by the defective genes are abnormal, causing the muscles to waste away. Unable to function properly, the muscle cells die and are replaced by fat and connective tissue. The symptoms of MD may not be noticed until as much as 50% of the muscle tissue has been affected.

DMD is X-linked, affects mostly males, and, according to the NCBI (2007, http://www.ncbi.nlm.nih.gov/ books/bv.fcgi?call=bv.View..ShowSection&rid=gnd.section .161), strikes 1 out of 3,500 boys worldwide. The gene for DMD is located on the X chromosome and encodes a large protein called dystrophin. (See Figure 5.17.)

Dystrophin provides structural support for muscle cells and without it the cell membrane becomes penetrable, allowing extracellular components into the cell. (See Figure 5.18.) These additional components increase the intracellular pressure, causing the muscle cell to die.

With myotonic dystrophy the muscles contract but have diminishing ability to relax, and there is muscle weakening and wasting. Typically, the initial complaints are the loss of hand strength or tripping while walking or climbing stairs. Along with decreased muscle strength, myotonic dystrophy may cause mental deficiency, hair loss, and cataracts. According to HealthAtoZ.com (2006, https://www.healthatoz .com/healthatoz/Atoz/common/standard/transform.jsp?request URI=/healthatoz/Atoz/ency/myotonic_dystrophy.jsp), it is an autosomal dominant disorder that occurs in 1 out of 20,000 people; it usually begins in young adulthood but can start at any age and varies in terms of severity.

The myotonic dystrophy gene is a protein kinase gene found on the long arm of chromosome 19. (See Figure 5.19.) The defect is a repeated set of three nucleotides—cytosine (C), thymine (T), and guanine (G)—in the gene. The symptoms of myotonic dystrophy frequently become more severe with each generation because mistakes in copying the gene from one generation to the next result in amplification of a genomic AGC/CTG triplet repeat, similar to the process observed in Huntington's disease. Unaffected individuals have CTG repeats with between three and thirty-seven iterations (repetitions) of the triplet. In contrast, people with the mild phenotype of myotonic dystrophy have between 40 and 170, and those with more serious forms of the disease have between 100 and 1,000 iterations.

All the various disorders labeled MD cause progressive weakening and wasting of muscle tissues. They vary, however, in terms of the usual age at the onset of symptoms, rate of progression, and initial group of muscles affected. The most common type, DMD, affects young boys, who show symptoms in early childhood and usually die from respiratory weakness or damage to the heart before adulthood. The gene is passed from the mother to her children. Females who inherit the defective gene generally do not manifest symptoms—they become carriers of the defective genes, and their children have a 50% chance of inheriting the disease. Other forms of MD appear later in life and are usually not fatal.

In 1992 scientists discovered the defect in the gene that causes myotonic dystrophy. In people with this disorder, a segment of the gene is enlarged and unstable. This finding helps physicians more accurately diagnose myotonic dystrophy. Researchers have since identified genes linked to other types of MD, including DMD, Becker MD, limb-girdle MD, and Emery-Dreifuss MD.

In 2005 Duygu Selcen and Andrew G. Engel of the Mayo Clinic identified a new form of MD that involves

FIGURE 5.17

Chromosome X and the DMD gene associated with Duchenne muscular dystrophy

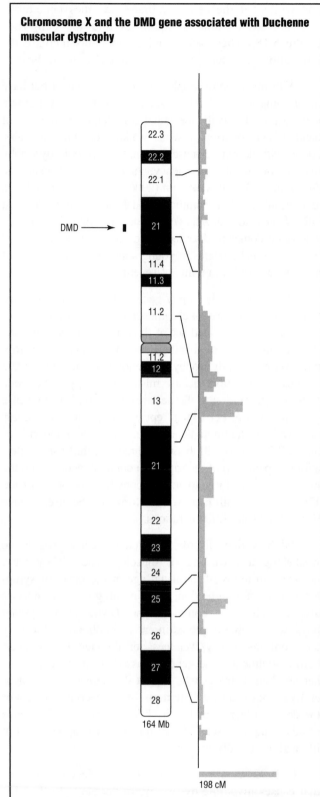

DMD

22.3
22.2
22.1
21
11.4
11.3
11.2
11.2
12
13
21
22
23
24
25
26
27
28

164 Mb

198 cM

SOURCE: "The DMD Gene Maps to Chromosome X," in "Duchenne Muscular Dystrophy," in *Science: The Human Gene Map, Genome Maps 7*, National Institutes of Health, U.S. National Library of Medicine, National Center for Biotechnology Information, 2004 http://www.ncbi.nlm.nih.gov/SCIENCE96/gene.cgi?DMD (accessed October 19, 2006)

FIGURE 5.18

Dystrophin and utrophin

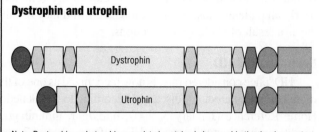

Dystrophin

Utrophin

Note: Dystrophin and utrophin are related proteins being used in the development of therapy for Duchenne muscular dystrophy (DMD).

SOURCE: Adapted from "Dystrophin and Utrophin," in *Genes and Disease*, National Institutes of Health, U.S. National Library of Medicine, National Center for Biotechnology Information, http://www.ncbi.nlm.nih.gov/books/bv.fcgi?rid=gnd.section.161 (accessed October 19, 2006)

FIGURE 5.19

Chromosome 19 and the DM gene associated with myotonic dystrophy

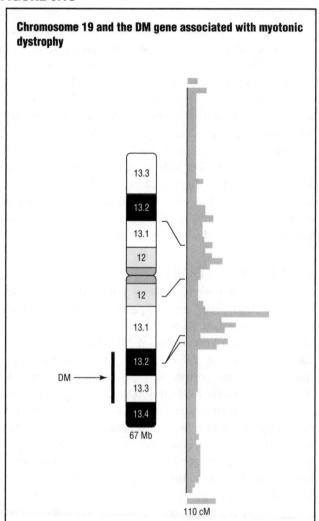

13.3
13.2
13.1
12
12
13.1
13.2
13.3
13.4

DM

67 Mb

110 cM

SOURCE: "The DM Gene Maps to Chromosome 19," in "Myotonic Dystrophy," in *Science: The Human Gene Map, Genome Maps 7*, National Institutes of Health, U.S. National Library of Medicine, National Center for Biotechnology Information, http://www.ncbi.nlm.nih.gov/SCIENCE96/gene.cgi?DM (accessed December 14, 2006)

Genetics and Genetic Engineering

mutations in a protein called ZASP, which binds to cardiac (heart) and skeletal muscles, and reported their finding in "Mutations in ZASP Define a Novel Form of Muscular Dystrophy in Humans" (*Annals of Neurology*, February 2005). Selcen and Engel detected ZASP mutations in eleven patients; in seven of these, they observed a dominant pattern of inheritance.

Treatment and Hope

There is no known cure for MD, but patients can be made more comfortable and functional by a combination of physical therapy, exercise programs, and orthopedic devices (special shoes, braces, or powered wheelchairs) that help them to maintain mobility and independence as long as possible.

Genetic research offers hope of finding effective treatments, and even cures, for these diseases. Gene therapy experiments designed to find a cure or a treatment for one or more of these types of MD are ongoing. Research teams have identified the crucial proteins produced by these genes, such as dystrophin, beta sarcoglycan, gamma sarcoglycan, and adhalin. One experimental treatment approach involves substituting a protein of comparable size, such as utrophin for dystrophin, to compensate for the loss of dystrophin. (See Figure 5.18.)

Because defective or absent proteins cause MD, researchers hope that experimental treatments to transplant normal muscle cells into wasting muscles will replace the diseased cells. Muscle cells, unlike other cells in the body, fuse together to become giant cells. Scientists hope that if cells with healthy genes can be introduced into the muscles and accepted by the body's immune system, the muscle cells will then begin to produce the missing proteins.

New delivery methods called vectors are also being tested, such as implanting a healthy gene into a virus that has been stripped of all of its harmful properties and then injecting the modified virus into a patient. Researchers hope this will reduce the amount of rejection by the patient's immune system, allowing the healthy gene to restore the missing muscle protein.

PHENYLKETONURIA

Phenylketonuria (PKU) is an example of a disorder caused by a gene-environment interaction. As a result of the defect, the affected individual is unable to convert phenylalanine into tyrosine. Phenylalanine in the body accumulates in the blood and can reach toxic levels. (See Figure 5.4.) This toxicity may impair brain and nerve development and result in mental retardation, organ damage, and unusual posture. When it occurs during pregnancy, it may jeopardize the health and viability of the unborn child.

Originally, PKU was considered simply an autosomal recessive inherited error of metabolism that occurred when an individual received two defective copies, caused by mutations in both alleles of the phenylalanine hydroxylase gene found on chromosome 12. (See Figure 5.20.) The environmental trigger—dietary phenylalanine—was not identified at first because phenylalanine is so prevalent in the diet, occurring in common foods such as milk and eggs and in the artificial sweetener aspartame. Recognition of dietary phenylalanine as a critical environmental trigger has enabled children born with PKU to lead normal lives when they are placed on low-phenylalanine diets, and mothers with the disease can bear healthy children.

The March of Dimes reports that to identify people at risk of PKU, all newborns in the United States are screened at birth for high levels of phenylalanine in the blood (http:// search.marchofdimes.com/cgi-bin/MsmGo.exe?grab_id=0 &page_id=162&query=PKU&hiword=PKU%20). After a second screening is done for those with elevated blood levels, approximately 1 out of 10,000 infants is diagnosed with PKU and, with proper diet, is likely to lead a healthy, normal life.

SICKLE-CELL DISEASE

Sickle-cell disease (SCD) is a group of hereditary diseases, including sickle-cell anemia (SCA) and sickle B-thalassemia, in which the red blood cells contain an abnormal hemoglobin, called hemoglobin S (HbS). HbS is responsible for the premature destruction of red blood cells, or hemolysis. In addition, it causes the red cells to become deformed, actually taking on a sickle shape, particularly in parts of the body where the amount of oxygen is relatively low. These abnormally shaped cells cannot travel smoothly through the smaller blood vessels and capillaries. They tend to clog the vessels and prevent blood from reaching vital tissues. This blockage produces anoxia (lack of oxygen), which in turn causes more sickling and more damage.

SCA is an autosomal recessive disease caused by a point mutation in the hemoglobin beta gene (HBB) found on chromosome 11p15.5. (Figure 5.2.) A mutation in HBB results in the production of hemoglobin with an abnormal structure. Figure 5.21 shows how a point mutation in SCA causes the amino acid glutamine to be replaced by valine to produce abnormal hemoglobin called HbS. It also shows how when red blood cells with HbS are oxygen-deprived they become sickle shaped and may cause blockages that result in tissue death.

Symptoms of SCA

People with SCA have symptoms of anemia, including fatigue, weakness, fainting, and palpitations or an increased awareness of their heartbeat. These palpitations

FIGURE 5.20

Chromosome 12 and the PAH gene associated with phenylketonuria (PKU)

13

12

11.2

12

13

14

15

21

22

23

PAH →

24.1

24.2

24.3

143 Mb

169 cM

SOURCE: "The PAH Gene Maps to Chromosome 12," in "Phenylketonuria (PKU)," in *Science: The Human Gene Map, Genome Maps 7*, National Institutes of Health, U.S. National Library of Medicine, National Center for Biotechnology Information, http://www.ncbi.nlm.nih.gov/SCIENCE96/gene.cgi?PAH (accessed October 19, 2006)

FIGURE 5.21

Sickle-cell anemia

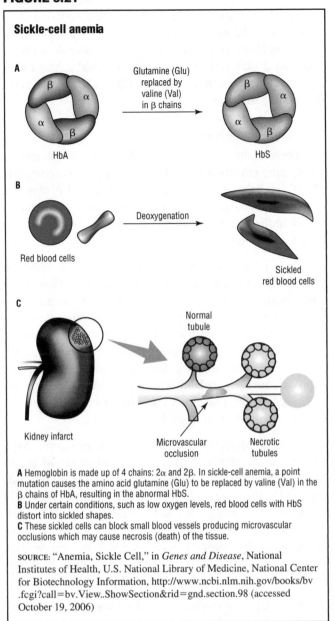

A Hemoglobin is made up of 4 chains: 2α and 2β. In sickle-cell anemia, a point mutation causes the amino acid glutamine (Glu) to be replaced by valine (Val) in the β chains of HbA, resulting in the abnormal HbS.
B Under certain conditions, such as low oxygen levels, red blood cells with HbS distort into sickled shapes.
C These sickled cells can block small blood vessels producing microvascular occlusions which may cause necrosis (death) of the tissue.

SOURCE: "Anemia, Sickle Cell," in *Genes and Disease*, National Institutes of Health, U.S. National Library of Medicine, National Center for Biotechnology Information, http://www.ncbi.nlm.nih.gov/books/bv.fcgi?call=bv.View..ShowSection&rid=gnd.section.98 (accessed October 19, 2006)

clots may also develop in the lungs, kidneys, brain, and other organs. A severe crisis or several acute crises can permanently damage various organs of the body. This damage can lead to death from heart failure, kidney failure, or stroke. The frequency of these crises varies from patient to patient. A sickle-cell crisis, however, occurs more often during infections and after an accident or an injury.

Who Contracts SCD?

Both the sickle-cell trait and the disease exist almost exclusively in people of African, Native American, and Hispanic descent and in those from parts of Italy, Greece, Middle Eastern countries, and India. If one parent has the sickle-cell gene, then the couple's offspring will carry the trait; if both the mother and the father have the trait, then

result from the heart's attempts to compensate for the anemia by pumping blood faster than normal.

In addition, patients experience occasional sickle-cell crises—attacks of pain in the bones and abdomen. Blood their children may be born with SCA. This trait is relatively common among African-Americans. People of African descent are advised to seek genetic counseling and testing for the trait before starting a family. According to the World Health Organization (March 10, 2006, http://www.who.int/genomics/public/geneticdiseases/en/index2.html#SCA), the sickle-cell trait is present in one out of twelve African-Americans, or about 2 million people. SCD is the most common inherited blood disorder in the United States, affecting 72,000 Americans, most of whom have African ancestry. SCA occurs in approximately 1 out of 500 African-American births and in 1 out of 1,000 to 1,400 Hispanic births. The occurrence of SCA in other groups is much lower.

Treatment of SCD

There is no universal cure for SCD, but the symptoms can be treated. Crises accompanied by extreme pain are the most common problems and can usually be treated with painkillers. Maintaining healthy eating and behavior and prompt treatment for any type of infection or injury is important. Special precautions are often necessary before any type of surgery, and for major surgery some patients receive transfusions to boost their levels of hemoglobin (the oxygen-bearing, iron-containing protein in red blood cells). In early 1995 a medication that prevented the cells from clogging vessels and cutting off oxygen was approved by the Food and Drug Administration.

BIOMEDICAL ADVANCES. Many adults with SCD now take hydroxyurea, an anticancer drug that causes the body to produce red blood cells that resist sickling. In 1995 a multicenter study showed that among adults with three or more painful crises per year, hydroxyurea lowered the median number of crises requiring hospitalization by 58%. In 2003 Martin H. Steinberg et al. reported in "Effect of Hydroxyurea on Mortality and Morbidity in Adult Sickle Cell Anemia: Risks and Benefits Up to 9 Years of Treatment" (*Journal of the American Medical Association*, April 2, 2003) that not only do patients on hydroxyurea have fewer crises but also they have a significant survival advantage when compared to SCD patients who do not take the medication. Subjects treated with it showed 40% lower mortality than others.

TAY-SACHS DISEASE

Tay-Sachs disease (TSD) is caused by mutations in the HEXA gene, located on the long arm of chromosome 15. (See Figure 5.22.) It is a fatal genetic disorder in children that causes the progressive destruction of the central nervous system. It is caused by the absence of an important enzyme called hexosaminidase A (hex-A). Without hex-A, a fatty substance called GM2 ganglioside builds up abnormally in the cells, particularly the brain's nerve cells. (Figure 5.23 shows how the absence of, or defect in, the hex-A protein prevents complete processing of GM2 ganglioside.) Eventually, these cells degenerate and die. This destructive process

FIGURE 5.22

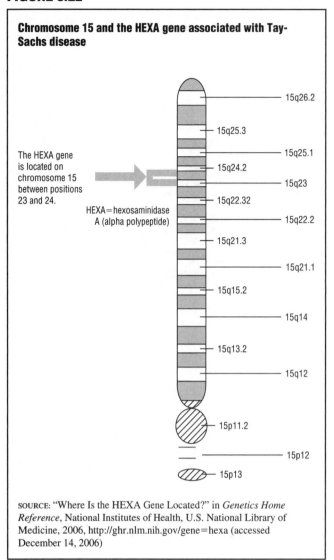

Chromosome 15 and the HEXA gene associated with Tay-Sachs disease

The HEXA gene is located on chromosome 15 between positions 23 and 24.

HEXA = hexosaminidase A (alpha polypeptide)

15q26.2
15q25.3
15q25.1
15q24.2
15q23
15q22.32
15q22.2
15q21.3
15q21.1
15q15.2
15q14
15q13.2
15q12
15p11.2
15p12
15p13

SOURCE: "Where Is the HEXA Gene Located?" in *Genetics Home Reference*, National Institutes of Health, U.S. National Library of Medicine, 2006, http://ghr.nlm.nih.gov/gene=hexa (accessed December 14, 2006)

begins early in the development of a fetus, but the disease is not usually diagnosed until the baby is several months old. By the time a child with TSD is four or five years old, the nervous system is so badly damaged that the child dies.

How Is TSD Inherited?

TSD is an autosomal recessive genetic disorder caused by mutations in both alleles of the HEXA gene on chromosome 15. Both the mother and the father must be carriers of the defective TSD gene to produce a child with the disease.

People who carry the gene for TSD are entirely unaffected and usually unaware that they have the potential to pass this disease to their offspring. A blood test distinguishes Tay-Sachs carriers from noncarriers. Blood samples may be analyzed by enzyme assay or DNA studies. Enzyme assay measures the level of hex-A in blood. Carriers have less hex-A than noncarriers. When only one parent is a carrier, the couple will not have a child with TSD. When both parents carry the recessive

FIGURE 5.23

Molecular basis for Tay-Sachs disease

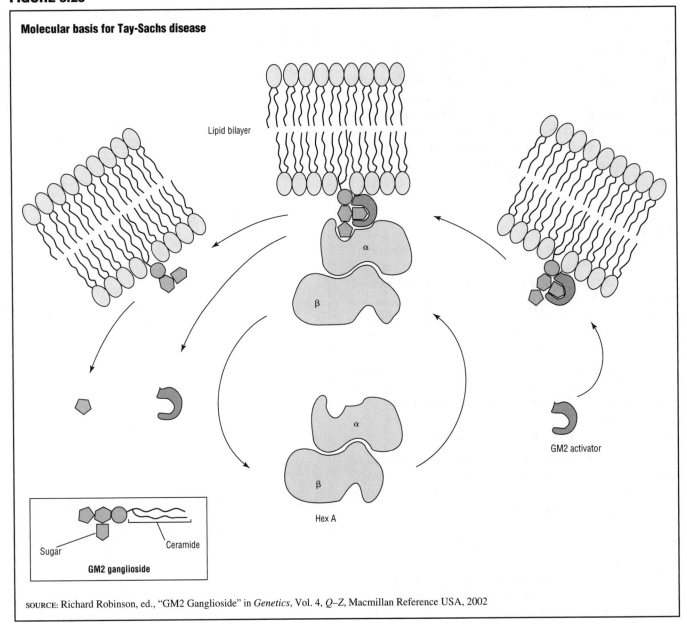

Lipid bilayer

GM2 activator

Hex A

Sugar

Ceramide

GM2 ganglioside

SOURCE: Richard Robinson, ed., "GM2 Ganglioside" in *Genetics*, Vol. 4, *Q–Z*, Macmillan Reference USA, 2002

TSD gene, they have a one in four chance in every pregnancy of having a child with the disease. They also have a 50% chance of bearing a child who is also a carrier. Prenatal diagnosis early in pregnancy can predict if the unborn child has TSD. If the fetus has the disease, the couple may choose to terminate the pregnancy.

Who Is at Risk?

Some genetic diseases, such as TSD, occur most frequently in a specific population. As the American neurologist Bernard Sachs observed, individuals of East European (Ashkenazi) Jewish descent have the highest risk of being carriers of TSD. According to the National Tay-Sachs and Allied Diseases Association (2007, http://www.tay-sachs.org/taysachs.php), approximately one out of twenty-seven Jews in the United States is a carrier of the TSD gene. French-Canadians and Cajuns also have the same carrier rate as Ashkenazi Jews. In the general population the carrier rate is 1 out of 250.

CHAPTER 6
GENETIC TESTING

"Diagnosis" means finding the cause *of a disorder, not just giving it a name.*

—Sydney Walker III, *A Dose of Sanity: Mind, Medicine, and Misdiagnosis* (1996)

Over the course of the last decade the definitions of health and disease have been transformed by advances in genetics. Genetic testing has enabled researchers and clinicians to detect inherited traits, diagnose heritable conditions, determine and quantify the likelihood that a heritable disease will develop, and identify genetic susceptibility to familial disorders. Many of the strides made in genetic diagnostics are direct results of the Human Genome Project, an international thirteen-year effort begun in 1990 by the U.S. Department of Energy and the National Institutes of Health, which mapped and sequenced the human genome in its entirety. The increasing availability of genetic testing has been one of the most immediate applications of this groundbreaking research.

A genetic test is the analysis of human deoxyribonucleic acid (DNA), ribonucleic acid (RNA), chromosomes, and proteins to detect heritable disease-related genotypes, mutations, phenotypes, or karyotypes (standard pictures of the chromosomes in a cell) for the purposes of diagnosis, treatment, and other clinical decision making. Most genetic testing is performed by drawing a blood sample and extracting DNA from white blood cells. Genetic tests may detect mutations at the chromosomal level, such as additional, absent, or rearranged chromosomal material, or even subtler abnormalities such as a substitution in one of the bases that make up the DNA. There is a broad range of techniques that can be used for genetic testing. Genetic tests have diverse purposes, including screening for and diagnosis of genetic disease in newborns, children, and adults; the identification of future health risks; the prediction of drug responses; and the assessment of risks to future children.

There is a difference between genetic tests performed to screen for disease and testing conducted to establish a diagnosis. Diagnostic tests are intended to definitively determine whether a patient has a particular problem. They are generally complex tests and commonly require sophisticated analysis and interpretation. They may be expensive and are generally performed only on people believed to be at risk, such as patients who already have symptoms of a specific disease.

In contrast, screening is performed on healthy, asymptomatic (showing no symptoms of disease) people and often to the entire relevant population. A good screening test is relatively inexpensive, easy to use and interpret, and helps identify which individuals in the population are at higher risk of developing a specific disease. By definition, screening tests identify people who need further testing or those who should take special preventive measures or precautions. For example, people who are found to be especially susceptible to genetic conditions with specific environmental triggers are advised to avoid the environmental factors linked to developing the disease. Examples of genetic tests used to screen relevant populations include those that screen people of Ashkenazi Jewish heritage (the East European Jewish population primarily from Germany, Poland, and Russia, as opposed to the Sephardic Jewish population primarily from Spain, parts of France, Italy, and North Africa) for Tay-Sachs disease, African-Americans for sickle-cell disease, and the fetuses of expectant mothers over age thirty-five for Down syndrome.

QUALITY AND UTILITY OF GENETIC TESTS

Like all diagnostic and screening tests, the quality and utility of genetic tests depend on their reliability, validity, sensitivity, specificity, positive predictive value, and negative predictive value. Reliability of testing refers to the test's ability to be repeated and to produce equivalent results in comparable circumstances. A reliable test is consistent and measures the same way each time it is used with the same patients in the same circumstances. For example, a well-calibrated balance scale is a reliable instrument for measuring body weight.

Validity is the accuracy of the test. It is the degree to which the test correctly identifies the presence of disease, blood level, or other quality or characteristic it is intended to detect. For example, if you put an object you knew weighed ten pounds on a scale and the scale said it weighed ten pounds, then the scale's results are valid. There are two components of validity: sensitivity and specificity.

Sensitivity is the test's ability to identify people who have the disease. Mathematically speaking, it is the percentage of people with the disease who test positive for the disease. Specificity is the test's ability to identify people who do not have the disease—it is the percentage of people without the disease who test negative for the disease. Ideally, diagnostic and screening tests should be highly sensitive and highly specific, thereby accurately classifying all people tested as either positive or negative. In practice, however, sensitivity and specificity are frequently inversely related—most tests with high levels of sensitivity have low specificity, and the reverse is also true.

The likelihood that a test result will be incorrect can be gauged based on the sensitivity and specificity of the test. For example, if a test's sensitivity is 95%, then when one hundred patients with the disease are tested, ninety-five will test positive and five will test false negative—they have the disease but the test has failed to detect it. For example, disorders such as Charcot-Marie-Tooth disease (a group of inherited, slowly progressive disorders that result from progressive damage to nerves; its symptoms include numbness and wasting of muscle tissue in the feet and legs, then in the hands and arms) can arise from mutations in one of many different genes, and because some of these genes have not yet been identified, they will not be detected and a false negative result might be reported. By contrast, if a test is 90% specific, when one hundred healthy, disease-free people are tested, ninety will receive negative test results and ten will be given false-positive results, meaning that they do not have the disease but the test has inaccurately classified them as positive.

The positive predictive value is the percentage of people that actually have the disease of all those with positive test results. The negative predictive value measures the percent of all the people with negative test results who do not have the disease.

PREGNANCY, CHILDBIRTH, AND GENETIC TESTING

There are thousands of genetic diseases, such as sickle-cell anemia, cystic fibrosis, and Tay-Sachs disease, that may be passed from one generation to the next.

Many tests have been developed to help screen parents at risk of passing on genetic disease to their children as well as to identify embryos, fetuses, and newborns who suffer from genetic diseases.

Carrier Identification

Carrier identification is the term for genetic testing to determine whether a healthy individual has a gene that may cause disease if passed on to his or her offspring. It is usually performed on people considered to be at higher than average risk, such as those of Ashkenazi Jewish descent, who have a 1 in 27 chance of being Tay-Sachs carriers (in other populations the risk is 1 out of 250), according to the National Tay-Sachs and Allied Diseases Association (2007, http://www.tay-sachs.org/taysachs.php). Testing is necessary because many carriers have just one copy of a gene for an autosomal recessive trait and are unaffected by the trait or disorder. Only someone with two copies of the gene will actually have the disorder. So while it is widely assumed that everyone is an unaffected carrier of at least one autosomal recessive gene, it only presents a problem in terms of inheritance when two parents have the same recessive disorder gene (or both are carriers). In this instance the offspring would each have a one in four chance of receiving a defective copy of the gene from each parent and developing the disorder. Figure 6.1 shows the pattern of inheritance of an autosomal recessive disorder.

Carrier testing is offered to individuals who have family members with a genetic condition, people with family members who are identified carriers, and members of racial and ethnic groups known to be at high risk. Figure 6.2 shows how carrier testing would be used for a family affected by cystic fibrosis and among African-Americans who may carry the gene for sickle-cell anemia.

Preimplantation Genetic Diagnosis

Preimplantation diagnosis is a newer genetic test that enables parents undergoing in vitro fertilization (fertilization that takes place outside the body) to screen an embryo for specific genetic mutations when it is no larger than six or eight cells and before it is implanted in the uterus to grow and develop. Figure 6.3 shows how preimplantation genetic diagnosis is performed.

Prenatal Testing

Prenatal genetic testing enables physicians to diagnose diseases in the fetus. Most genetic tests examine blood or other tissue to detect abnormalities. An example of a blood test is the triple marker screen. This test measures levels of alpha fetoprotein (AFP), human chorionic gonadotropin (hCG), and unconjugated estriol and can identify some birth defects such as Down syndrome and neural tube defects. (Two of the most common neural tube defects are anencephaly—absence of most of the

FIGURE 6.1

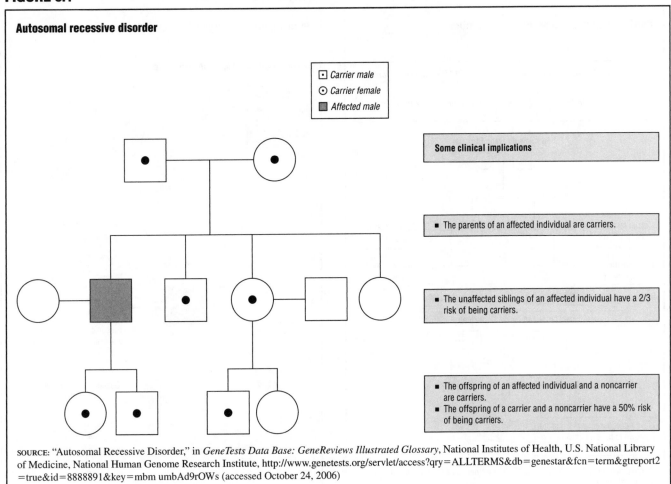

Autosomal recessive disorder

☑ *Carrier male*
☉ *Carrier female*
■ *Affected male*

Some clinical implications

- The parents of an affected individual are carriers.

- The unaffected siblings of an affected individual have a 2/3 risk of being carriers.

- The offspring of an affected individual and a noncarrier are carriers.
- The offspring of a carrier and a noncarrier have a 50% risk of being carriers.

SOURCE: "Autosomal Recessive Disorder," in *GeneTests Data Base: GeneReviews Illustrated Glossary*, National Institutes of Health, U.S. National Library of Medicine, National Human Genome Research Institute, http://www.genetests.org/servlet/access?qry=ALLTERMS&db=genestar&fcn=term&gtreport2=true&id=8888891&key=mbm umbAd9rOWs (accessed October 24, 2006)

brain—and spina bifida—incomplete development of the back and spine.)

The fetal yolk sac and the fetal liver make AFP, which is continuously processed by the fetus and excreted into the amniotic fluid. A small amount crosses the placenta and can be found in maternal blood. Maternal screening for AFP levels is based on maternal age, fetal gestation, and the number of fetuses the mother is carrying. Elevated levels of AFP are associated with conditions such as spina bifida and low levels are found with Down syndrome. Because AFP levels alone may not always adequately detect disorders, two other blood serum tests have been developed. hCG is a glycoprotein produced by the placenta. Normally, hCG is elevated at the time of implantation, but decreases at about eight weeks of gestation, and then drops again at approximately twelve weeks of gestation. Elevated levels of hCG are found with Down syndrome. The placenta also produces unconjugated estriol. As with AFP, lower unconjugated estriol maternal serum levels are also found with Down syndrome. Triple marker screen results are usually available within several days and women with abnormal results are often advised to undergo additional diagnostic testing such as chorionic villus sampling

(CVS), amniocentesis, or percutaneous umbilical blood sampling (withdrawing blood from the umbilical cord).

CVS enables obstetricians and perinatologists (physicians specializing in evaluation and care of high-risk expectant mothers and infants) to assess the progress of pregnancy during the first trimester (the first three months). A physician passes a small, flexible tube called a catheter through the cervix to extract chorionic villi tissue—cells that will become the placenta and are genetically identical to the baby's cells. The chorion develops from trophoblasts, or the same cells as the fetus, and contains the same DNA and chromosomes. The cells obtained via CVS are examined in the laboratory for indications of genetic disorders such as cystic fibrosis, Down syndrome, Tay-Sachs, and thalassemia, and the results of testing are available within seven to fourteen days. Table 6.1 describes CVS and other prenatal diagnostic tests and specifies when during pregnancy they are performed.

Amniocentesis involves taking a sample of the fluid that surrounds the fetus in the uterus for chromosome analysis. An amniocentesis is usually performed at fifteen to twenty weeks of gestation, although it can be done as

FIGURE 6.2

Carrier testing for recessive gene mutation

□ Male: unaffected   ○ Female: unaffected   ■ Male: with cystic fibrosis

⊡/⊙ } Identified carrier   ▢/○ } Unaffected; not yet tested   ◇P Pregnant

Family with cystic fibrosis

Carrier testing is appropriate for:

A = individual with an affected family member

Carrier testing is appropriate for:

B = family member of an identified carrier

African-American   African-American

Carrier testing is appropriate for:

C = individuals in a racial group known to have a higher carrier rate for a particular condition

Approximately one in ten individuals of African-American ancestry is a carrier for sickle-cell anemia.

SOURCE: "Carrier Testing," in *GeneTests Data Base: GeneReviews Illustrated Glossary*, National Institutes of Health, U.S. National Library of Medicine, National Human Genome Research Institute, http://www.genetests.org/servlet/access?qry=ALLTERMS&db= genestar&fcn=term&gtreport2=true&id=8888891&key=mbmumb Ad9rOWs (accessed October 26, 2006)

FIGURE 6.3

Preimplantation diagnosis of genetic disorders in embryos

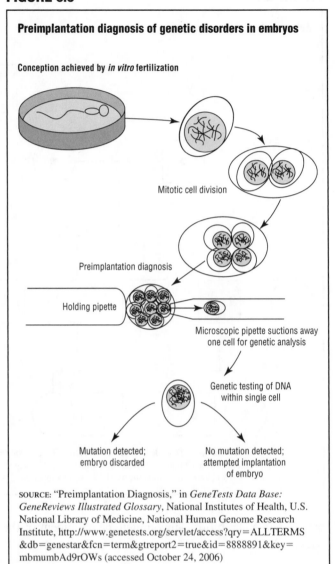

Conception achieved by *in vitro* fertilization

Mitotic cell division

Preimplantation diagnosis

Holding pipette

Microscopic pipette suctions away one cell for genetic analysis

Genetic testing of DNA within single cell

Mutation detected; embryo discarded

No mutation detected; attempted implantation of embryo

SOURCE: "Preimplantation Diagnosis," in *GeneTests Data Base: GeneReviews Illustrated Glossary*, National Institutes of Health, U.S. National Library of Medicine, National Human Genome Research Institute, http://www.genetests.org/servlet/access?qry=ALLTERMS &db=genestar&fcn=term&gtreport2=true&id=8888891&key= mbmumbAd9rOWs (accessed October 24, 2006)

early as twelve weeks. (See the description of early amniocentesis in Table 6.1.) About twenty milliliters of amniotic fluid is obtained when the physician inserts a hollow needle through the abdominal wall and the wall of the uterus. Fetal karyotyping, DNA analysis, and biochemical testing may be performed on the isolated fetal cells. Like CVS, amniocentesis samples and analyzes cells derived from the baby to enable parents to learn of chromosomal abnormalities, as well as the gender of the unborn child, about two weeks after the test is performed.

Using samples of genetic material obtained from amniocentesis or CVS, physicians can detect disease in an unborn child. Down syndrome (also known as trisomy 21, because it is caused by an extra copy of chromosome 21) is the genetic disease most often identified using this technique. Down syndrome is rarely inherited; most cases result from an error in the formation of the ovum (egg) or sperm, leading to the inclusion of an extra chromosome

21 at conception. As with prenatal diagnosis for most inherited genetic diseases, this use of genetic testing is focused on reproductive decision making.

The most invasive prenatal procedure for genetic testing is periumbilical blood sampling. Table 6.1 describes the technique. Periumbilical blood sampling poses the greatest risk to the unborn child—one in fifty miscarriages occurs as a result of this procedure. It is used when a diagnosis must be made quickly. For example, when an expectant mother is exposed to an infectious agent with the potential to produce birth defects, it may be used to examine fetal blood for the presence of infection.

Until 2006 it was thought that women undergoing CVS were more likely to miscarry than those who had amniocentesis; however, Aaron B. Caugey, Linda M. Hopkins, and Mary E. Norton, in "Chorionic Villus Sampling Compared with Amniocentesis and the Difference in the Rate of Pregnancy Loss" (*Obstetrics and Gynecology*, September 2006), refute this notion. Caugey,

TABLE 6.1

Prenatal diagnostic testing methods

Procedure	Technique (ultrasound guided)	Sample	Timing in gestation*
Chorionic villus sampling (CVS)	Needle inserted through mother's abdomen or catheter through cervix	Chorionic villus	10–12 weeks
Early amniocentesis	Needle inserted through mother's abdomen into amniotic sac	Amniotic fluid and/or amniocytes	<15 weeks
Amniocentesis	Needle inserted through mother's abdomen into amniotic sac	Amniotic fluid and/or amniocytes	15–20 weeks
Placental biopsy	Needle inserted through mother's abdomen into placenta	Placental tissue	>12 weeks
Periumbilical blood sampling (PUBS) (aka cordocentesis)	Needle inserted through mother's abdomen into fetal umbilical vein	Fetal blood	>18 weeks
Fetoscopy with fetal skin biopsy	Needle inserted through mother's abdomen, camera used to facilitate biopsy	Fetal skin	>18 weeks

*Menstrual weeks calculated either from the first day of the last normal menstrual period or by ultrasound measurements.

SOURCE: "Prenatal Diagnosis," in *GeneTests Data Base: GeneReviews Illustrated Glossary*, National Institutes of Health, U.S. National Library of Medicine, National Human Genome Research Institute, http://www.genetests.org/servlet/access?qry=ALLTERMS&db=genestar&fcn=term&gtreport2=true&id=08888891&key=mbmumbAd9rOWs (accessed October 24,2006)

Hopkins, and Norton analyzed the outcomes of nearly 10,000 CVS and 32,000 amniocentesis procedures and found that CVS was no more likely than amniocentesis to lead to pregnancy loss. They attribute previously reported higher rates of miscarriage resulting from CVS to the fact that clinicians were not yet experienced at performing the newer procedure.

Genetic Testing of Newborns

The most common form of genetic testing is the screening of blood taken from newborn infants for genetic abnormalities. The Maternal and Child Health Bureau (September 2004, http://mchb.hrsa.gov/programs/genetics/presentations/comments/meeting2comments.htm) reports that in the United States more than four million newborns are screened every year for specific genetic disorders, such as phenylketonuria (PKU), and other medical conditions that are only indirectly genetically linked, such as congenital hypothyroidism (underactive thyroid gland). PKU is an inherited error of metabolism resulting from a deficiency of an enzyme called phenylalanine hydroxylase. The lack of this enzyme can produce mental retardation, organ damage, and postural problems. Children born with PKU must pay close attention to their diets to lead healthy, normal lives.

Laboratory Techniques for Genetic Prenatal Testing

Genetic testing is performed on chromosomes, genes, or gene products to determine whether a mutation is causing or may cause a specific condition. Direct testing examines the DNA or RNA that makes up a gene. Linkage testing looks for disease-causing gene markers in family members from at least two generations. Biochemical testing assays certain enzymes or proteins, which are the products of genes. Cytogenetic testing examines the chromosomes.

Generally, a blood sample or buccal smear (cells from the mouth) is used for genetic tests. Other tissues used include skin cells from a biopsy, fetal cells, or stored tissue samples. (Table 6.1 lists the tissue samples used in each procedure.) Testing requires highly trained, certified technicians and laboratories because the procedures are complex and varied, the technology is new and evolving, and hereditary conditions are often rare, so many testing techniques require special expertise. In the United States laboratories performing clinical genetic tests must be approved under the Clinical Laboratory Improvement Amendments (CLIA), passed by Congress in 1988 to establish the standards with which all laboratories that test human specimens must comply to receive certification. CLIA standards determine the qualifications of laboratory personnel, categorize the complexity of various tests, and oversee quality improvement and assurance. In 2005 about 600 laboratories in the United States were performing genetic testing to detect and diagnose more than 1,200 conditions. Figure 6.4 shows the tremendous growth of both laboratories and diseases for which testing is available from 1993 to 2005.

The polymerase chain reaction (PCR) technique permits rapid cloning and DNA analysis and allows selective amplification of specific DNA sequences. A polymerase chain reaction can be performed in hours and is a sensitive test that may be used to screen for altered genes, but it is limited by the size and length of the DNA sequences that can be cloned. Figure 6.5 shows how PCR is performed.

Fluorescence in situ hybridization (FISH), in which a fluorescent label is attached to a DNA probe that will bind to the complementary DNA strands, provides a unique opportunity to view specific genetic codes. The FISH technique may be used on cells and fluid obtained by CVS and amniocentesis as well as on maternal blood. With this technique, a single strand of DNA is used to create a probe that attaches at the specific gene location. To separate the double-stranded DNA, heat or chemicals are used to break the chemical bonds of the DNA and obtain a single strand. (See Figure 6.6.)

There are three kinds of chromosome-specific probes: repetitive probes, painting probes, and locus-specific probes. Repetitive probes produce intense signals by creating tandem

FIGURE 6.4

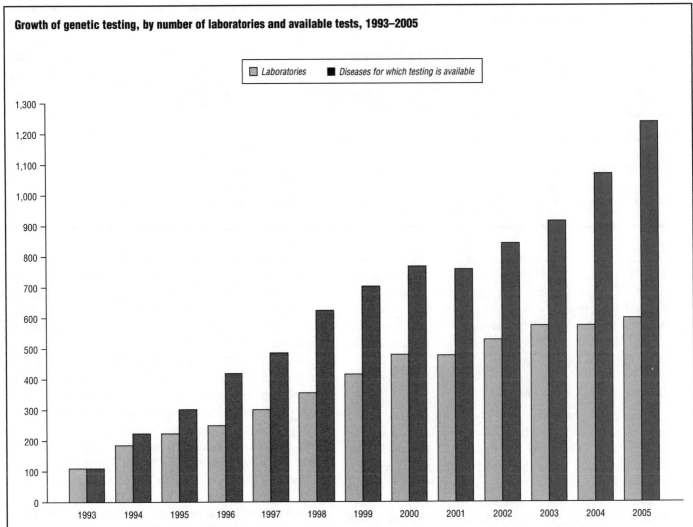

Growth of genetic testing, by number of laboratories and available tests, 1993–2005

Laboratories ■ Diseases for which testing is available

SOURCE: "Growth of Laboratory Directory," in *GeneTests Data Base: Laboratory Directory*, National Institutes of Health, U.S. National Library of Medicine, National Human Genome Research Institute, http://www.genetests.org/servlet/access?id=8888892&key=EesVvzHdDIGwZ&fcn=y&fw=u2c6&filename=/whatsnew/labdirgrowth.html (accessed December 14, 2006). Data from GeneTests database (2005), www.genetests.com.

repeats of base pairs. Painting probes are collections of specific DNA sequences that may extend along either part or all of an individual chromosome. These probe labels are most useful for identifying complex rearrangements of genetic material in structurally abnormal chromosomes. Probes that can hybridize to a single gene locus are called locus-specific probes. They can be used to identify a gene in a particular region of a chromosome. Locus-specific probes are used to identify a deletion or duplication of genetic material. The FISH procedure allows a signal to be visualized that indicates the presence or absence of DNA. The entire process can be completed in fewer than eight hours using five to seven probes. For this reason, FISH is frequently used as a rapid screen for trisomies and genetic disorders.

Although FISH is the most commonly used and most readily available prenatal diagnostic cytogenetic technique, it has limited ability to detect translocations, deletions, and inversions. Microdissection FISH (also used for prenatal diagnosis) is another method that is more sensitive to these

alterations. Microdissection FISH constructs probes to define specific regions of the human chromosome. Figure 6.7 shows how the microdissection technique is used to identify structurally abnormal chromosomes.

Newer technologies allow for the identification of all human chromosomes, thereby expanding the FISH application to the entire genome. Multiplex FISH applies combinations of probes to color components of fluorescent dye during metaphase to visualize each chromosome. This type of FISH procedure evolved from the whole-chromosome painting probes and uses a combination of fluorochromes to identify each chromosome. Just five fluorophores are needed to decode the entire complement of human chromosomes. Multicolor spectral karyotyping uses computer imaging and Fourier spectroscopy to increase the analysis of genetic markers. (See Figure 6.8.) Although FISH techniques are considered highly reliable, there are limitations to multiplex and multicolor FISH, in that inversion or subtle deletions may be overlooked. The greatest advant-

FIGURE 6.5

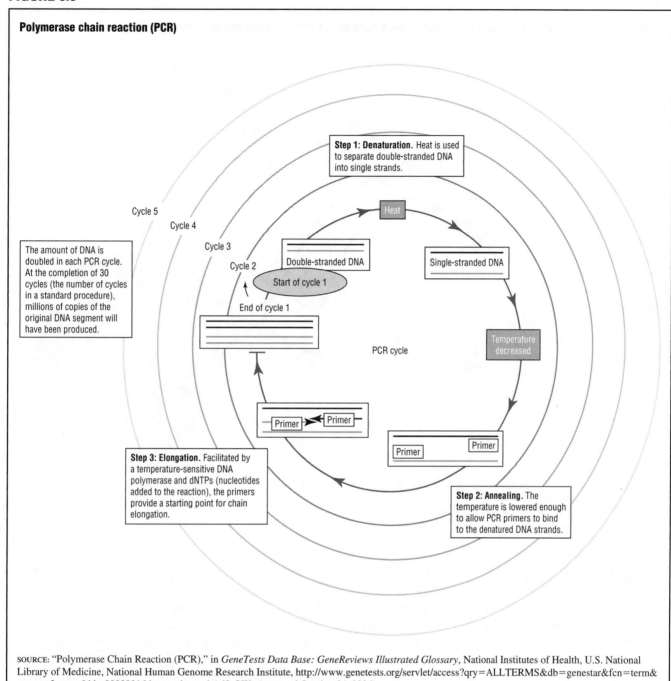

Polymerase chain reaction (PCR)

Step 1: Denaturation. Heat is used to separate double-stranded DNA into single strands.

Heat

Cycle 5

Cycle 4

Cycle 3

Cycle 2

Double-stranded DNA

Single-stranded DNA

Start of cycle 1

The amount of DNA is doubled in each PCR cycle. At the completion of 30 cycles (the number of cycles in a standard procedure), millions of copies of the original DNA segment will have been produced.

End of cycle 1

Temperature decreased

PCR cycle

Primer   Primer

Primer   Primer

Step 3: Elongation. Facilitated by a temperature-sensitive DNA polymerase and dNTPs (nucleotides added to the reaction), the primers provide a starting point for chain elongation.

Step 2: Annealing. The temperature is lowered enough to allow PCR primers to bind to the denatured DNA strands.

SOURCE: "Polymerase Chain Reaction (PCR)," in *GeneTests Data Base: GeneReviews Illustrated Glossary*, National Institutes of Health, U.S. National Library of Medicine, National Human Genome Research Institute, http://www.genetests.org/servlet/access?qry=ALLTERMS&db=genestar&fcn=term& gtreport2=true&id=8888891&key=mbmumbAd9rOWs (accessed October 24, 2006)

age of multiplex FISH is its application in cases of dysmorphic infants. Spectral karyotyping techniques can detect cryptic unbalanced translocations that cannot be identified by other techniques.

GENETIC TESTING FOR SICKLE-CELL ANEMIA. Sickle-cell anemia is an autosomal recessive disease that results when hemoglobin S is inherited from both parents. (When hemoglobin S is inherited from only one parent, the individual is a sickle-cell carrier.) Because normal and sickle hemoglobins differ at only one amino acid in

the hemoglobin gene, a test called hemoglobin electrophoresis is used to establish the diagnosis.

Genetic testing for sickle-cell anemia involves restriction fragment length polymorphism (DNA sequence variant) that uses specific enzymes to cut the DNA. (See Figure 6.9.) These enzymes cut the DNA at a specific base sequence on the normal gene but not on a gene in which a mutation is present. As a result of this technique, there are longer fragments of sickle hemoglobin. Another technique known as gel electrophoresis sorts

FIGURE 6.6

Fluorescence in situ hybridization (FISH) is useful for gene mapping and for identifying abnormalities in chromosomes

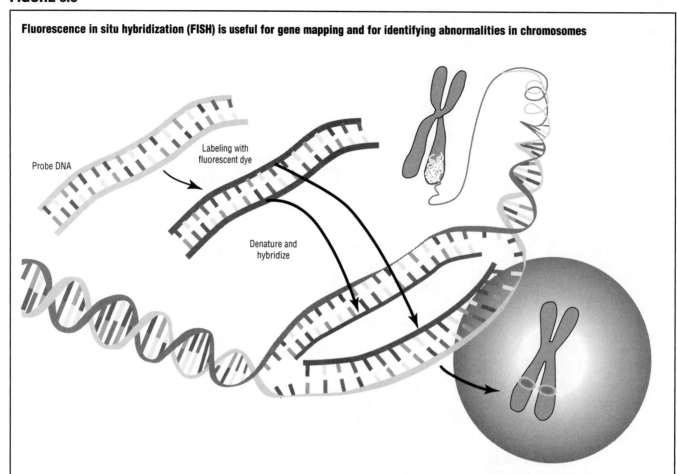

SOURCE: "Fluorescence in Situ Hybridization," in *Talking Glossary of Genetic Terms*, U.S. Department of Health and Human Services, National Institutes of Health, National Human Genome Research Institute, Division of Intramural Research, http://www.genome.gov/glossary.cfm?key=fluorescence%20in% 20 situ%20hybridization%20%28FISH%29 (accessed December 14, 2006)

the DNA fragments by size. Autoradiography renders the DNA fragments by generating an image after radioactive probes have labeled the DNA fragments that contain the specific gene sequence. The location of the fragments distinguishes carrier status (heterozygous) from sickle-cell anemia (homozygous), or normal blood.

New Techniques Detect Fetal Gene Mutations

In 2005 Ying Li et al. announced success with a technique that identifies fetal gene mutations such as beta-thalassemia from samples of maternal blood and reported their discovery in "Detection of Paternally Inherited Fetal Point Mutations for Beta-Thalassemia Using Size-Fractionated Cell-Free DNA in Maternal Plasma" (*Journal of the American Medical Association*, February 16, 2005). The technique relies on the fact that circulatory fetal DNA sequences comprise fewer than 300 base pairs, whereas maternal DNA exceeds 500 base pairs. Li and the other researchers used PCR amplification to select for paternally inherited DNA sequences. Presence of the paternal mutant alleles for beta-thalassemia was then detected by allele-specific PCR.

They tested the new technique on maternal blood samples from expectant mothers whose fetuses were at risk for beta-thalassemia because the father was a carrier for one of four beta-globin gene mutations. The results were verified by comparing the findings from chorionic villus sampling. Li et al. also reported the cost-effectiveness of the new technique. Because it does not require complex machinery and relies on currently available technology, they estimate the cost of a single analysis might be just $8.

Another new technique reported in 2005 was analysis of amniotic fluid using oligonucleotide hybridization and microarray data analysis. (Figure 6.10 shows how microarray technology is performed.) In "Global Gene Expression Analysis of the Living Human Fetus Using Cell-Free Messenger RNA in Amniotic Fluid" (*Journal of the American Medical Association*, February 16, 2005), Paige B. Larrabee et al. note that they found that gene expression patterns correlated with the gender of the fetus, gestational age, and disease status. Larrabee and her collaborators assert that this technology could assist

FIGURE 6.7

Microdissection

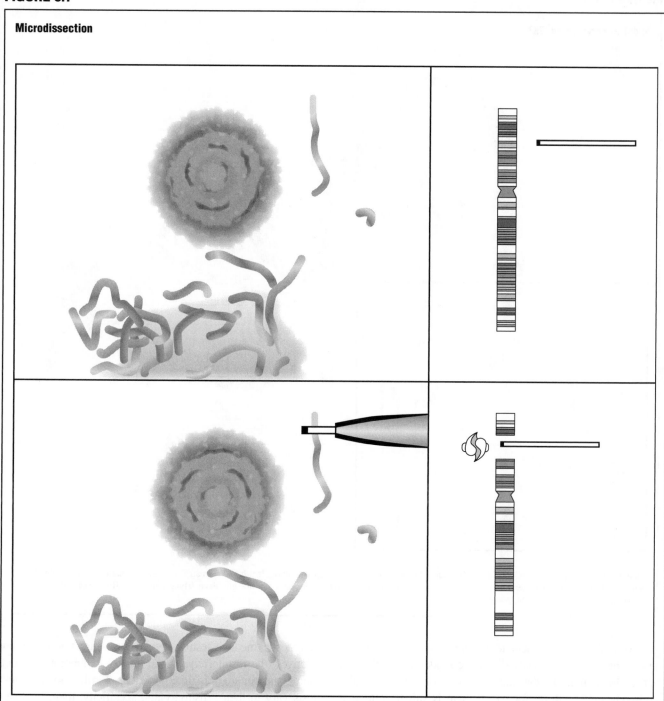

SOURCE: "How Does Microdissection Work?" in "Chromosome Microdissection Fact Sheet," in *Human Genome Project Information*, U.S. Department of Energy Office of Science, Office of Biological and Environmental Research, Human Genome Project, http://www.genome.gov/10000204 (accessed December 14, 2006)

in advancing human developmental research and in identifying new biomarkers for prenatal assessment.

GENETIC DIAGNOSIS IN CHILDREN AND ADULTS

Genetic testing can also be performed postnatally (after birth) to determine which children and adults are

at increased risk of developing specific diseases. By 2006 scientists could perform predictive genetic testing to identify which individuals were at risk for cystic fibrosis, Tay-Sachs disease, Huntington's disease, amyotrophic lateral sclerosis (a degenerative neurological condition commonly known as Lou Gehrig's disease), and several types of cancers, including some cases of breast, colon, and ovarian cancer.

FIGURE 6.8

Spectral karyotyping (SKY)

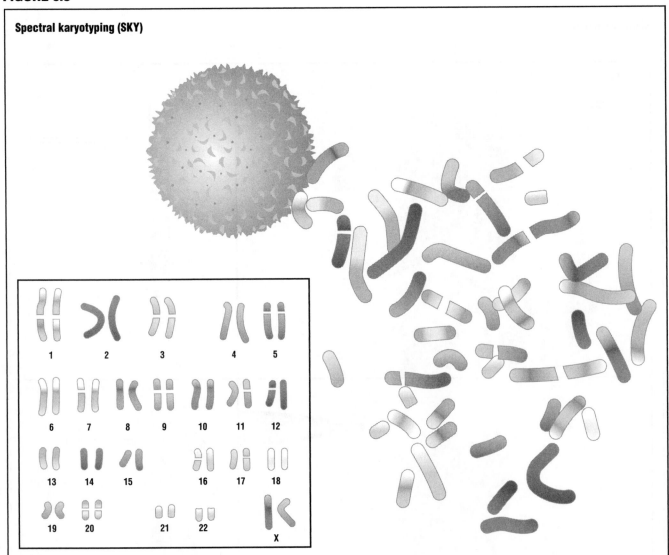

SOURCE: "Spectral Karyotyping," in *Talking Glossary of Genetic Terms*, U.S. Department of Health and Human Services, National Institutes of Health, National Human Genome Research Institute, Division of Intramural Research, http://www.genome.gov/Pages/Hyperion//DIR/VIP/Glossary/Illustration/sky.shtml (accessed October 24, 2006)

More than 1,200 genetic tests were available in 2006, but public health professionals did not consider it practical to screen for conditions that are rare, have only minor health consequences, or those for which there is still no effective treatment. The most frequently performed genetic tests were those considered most useful in terms of their potential to screen populations for diseases that occur relatively frequently, have serious medical consequences (including death) if untreated, and for which effective treatment is available.

A positive test result (the presence of mutation—a defective or altered gene) from predictive genetic testing does not guarantee that the individual will develop the disease; it simply identifies the individual as genetically susceptible and at an increased risk for developing the disease. For example, a woman who tests positive for the BRCA1 gene has about an 80% chance of developing breast cancer before age sixty. It is also important to note that, like other types of

diagnostic medical testing, genetic tests are not 100% predictive—the results rely on the quality of laboratory procedures and accuracy of interpretations. Furthermore, because tests vary in their sensitivity and specificity, there is always the possibility of false-positive and false-negative test results.

Researchers hope that positive test results will encourage people who are at higher than average risk of developing a disease to be especially vigilant about disease prevention and screening for early detection, when many diseases are most successfully treated. There is an expectation that genetic information will increasingly be used in routine population screening to determine individual susceptibility to common disorders such as heart disease, diabetes, and cancer. This type of screening will identify groups at risk so that primary prevention efforts such as diet and exercise or secondary prevention efforts such as early detection can be initiated.

FIGURE 6.9

Cutting DNA with restrictive enzymes

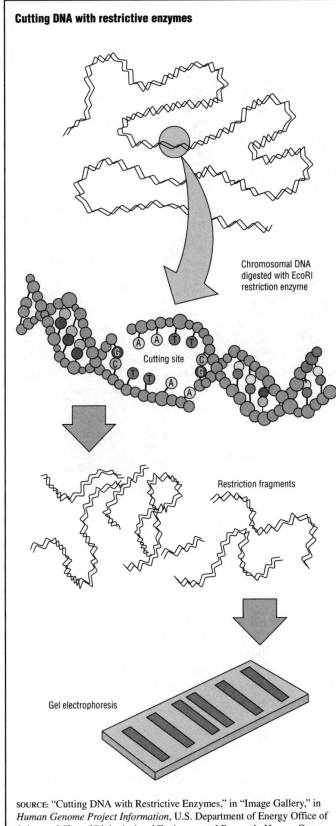

Chromosomal DNA digested with EcoRI restriction enzyme

Cutting site

Restriction fragments

Gel electrophoresis

SOURCE: "Cutting DNA with Restrictive Enzymes," in "Image Gallery," in *Human Genome Project Information*, U.S. Department of Energy Office of Science, Office of Biological and Environmental Research, Human Genome Project, http://www.ornl.gov/sci/techresources/Human_Genome/graphics/slides/ttenzymecut.shtml (accessed December 15, 2006)

Diagnostic Genetic Testing

Most genetic testing is performed on people who are asymptomatic (people who are apparently healthy). The objective of screening is to determine if people are carriers of a genetic disease or to identify their susceptibility or risk of developing a specific disease or disorder. There is, however, some testing performed on people with symptoms of a disease to clarify or establish the diagnosis and calculate the risk of developing the disease for other family members. This type of testing is known as diagnostic genetic testing or symptomatic genetic testing. It may also assist in directing treatment for symptomatic patients in whom a mutation in a single gene (or in a gene pair) accounts for a disorder. Cystic fibrosis and myotonic dystrophy are examples of disorders that may be confirmed or ruled out by diagnostic genetic testing and other methods (such as the sweat test for cystic fibrosis or a neurological evaluation for myotonic dystrophy).

One issue involved in diagnostic genetic testing is the appropriate frequency of testing in view of rapidly expanding genetic knowledge and identification of genes linked to disease. Physicians frequently see symptomatic patients for whom there is neither a definitive diagnosis nor a genetic test. The as-yet-unanswered question is: Should such people be recalled for genetic testing each time a new test becomes available? Although clinics and physicians who perform genetic testing counsel patients to maintain regular contact so they may learn about the availability of new tests, there is no uniform guideline or recommendation about the frequency of testing.

New Genetic Tests Help Patients Choose Optimal Treatment

In August 2005 the Food and Drug Administration (FDA) approved marketing of a new genetic test that will help physicians make personalized drug treatment decisions for some patients ("FDA Clears Genetic Test That Advances Personalized Medicine: Test Helps Determine Safety of Drug Therapy," August 22, 2005, http://www.fda.gov/bbs/topics/NEWS/2005/NEW01220.html). The Invader UGT1A1 Molecular Assay detects variations in a gene that affects how certain drugs are broken down and cleared by the body. Using this information, physicians can determine the optimal drug dosage for each patient and minimize the harsh and potentially life-threatening side effects of drug treatment. The new test joins a growing list of genetic tests used to customize treatment decisions, including the Roche AmpliChip, used to personalize dosage of antidepressants, antipsychotics, beta-blockers, and some chemotherapy drugs, and TRUGENE HIV-1 Genotyping Kit, used to detect variations in the genome of the human immunodeficiency virus (which can cause acquired immunodeficiency syndrome) that make the virus resistant to some antiretroviral drugs.

FIGURE 6.10

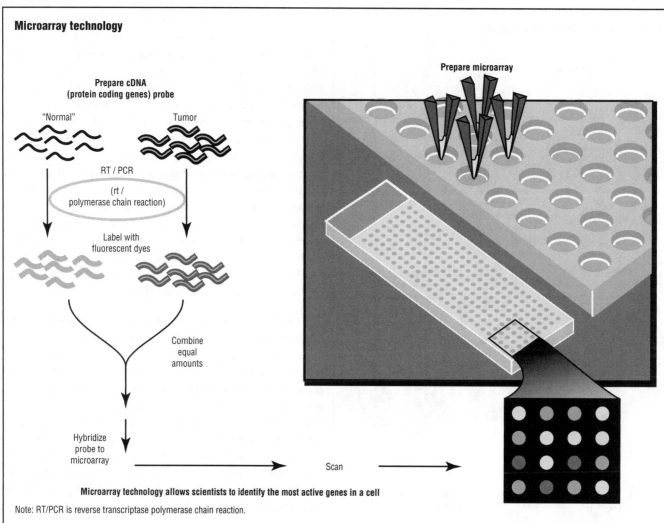

Microarray technology

Prepare cDNA
(protein coding genes) probe

"Normal"   Tumor

RT / PCR

(rt /
polymerase chain reaction)

Label with
fluorescent dyes

Combine
equal
amounts

Hybridize
probe to
microarray   Scan

Prepare microarray

Microarray technology allows scientists to identify the most active genes in a cell

Note: RT/PCR is reverse transcriptase polymerase chain reaction.

SOURCE: Adapted from "Microarray Technology," in *Talking Glossary of Genetic Terms*, U.S. Department of Health and Human Services, National Institutes of Health, National Human Genome Research Institute, Division of Intramural Research, http://www.genome.gov//Pages/Hyperion//DIR/VIP/Glossary/Illustration/microarray_technology.cfm (accessed December 15, 2006)

In 2006 Anil Potti et al. reported in "A Genomic Strategy to Refine Prognosis in Early-Stage Non-Small-Cell Lung Cancer" (*New England Journal of Medicine*, August 10, 2006) about a new genetic test that predicts with up to 90% accuracy which early-stage lung cancer patients are likely to experience recurrence and therefore would benefit from chemotherapy (drug treatment for cancer). Currently, patients diagnosed with early-stage non-small-cell lung cancer undergo surgery followed by observation, without chemotherapy. About one-third of such patients experience a relapse. Potti and his coauthors describe the test as "a fingerprint unique to the individual patient [that] predicts survival changes."

The National Cancer Institute announced in 2006 another new genetic test that was in clinical trials and that promised to distinguish which breast cancer patients can safely skip chemotherapy (May 23, 2006, http://www.cancer.gov/clinicaltrials/digestpage/TAILORx). The Onco-

type DX test examines the surgically removed tumor for twenty-one different genes whose interactions can predict the likelihood of a relapse and calculates the odds of a relapse on a scale from 0 to 100. A score greater than 30 indicates the benefits of chemotherapy, whereas a score lower than 18 suggests forgoing chemotherapy.

Population Screening

Population screening for heritable diseases is one potentially lifesaving application of molecular genetics technology. Prenatal screening has demonstrated benefits and gained widespread use; however, genetic screening has not yet become part of routine medical practice for adults. Geneticists have identified at least seven genes that might be candidates for use as population screening tests in adults in the United States. The genes include HFE, for hereditary hemochromatosis; apolipoprotein E-4, linked to Alzheimer's disease; CYP2D6, linked to ankylosing spondylitis (arthritis of the spine); BRCA1 and BRCA2, genes for hereditary breast and ovarian cancer; familial adenomatous polyposis, associated with precancerous growths in the colon; and factor V Leiden, the most common hereditary blood clotting disorder in the United States. As of 2007, screening for variants in these genes had not entered into routine medical practice because there was considerable controversy about the predictive value of testing for these genes and how to monitor and care for people who test positive for them.

Genetic Susceptibility

Susceptibility testing, also known as predictive testing, determines the likelihood that a healthy person with a family history of a disorder will develop the disease. Testing positive for a specific genetic mutation indicates an increased susceptibility to the disorder but does not establish a diagnosis. For example, a woman may choose to undergo testing to find out whether she has genetic mutations that would indicate likelihood of developing hereditary cancer of the breast or ovary. If she tests positive for the genetic mutation, she may then decide to undergo some form of preventive treatment. Preventive measures may include increased surveillance such as more frequent mammography, chemoprevention—prescription drug therapy intended to reduce risk—or surgical prophylaxis, such as mastectomy and/or oophorectomy (surgical removal of the breasts and ovaries, respectively).

Testing Children for Adult-Onset Disorders

In 2000 the American Academy of Pediatrics Committee on Genetics recommended genetic testing for people under age eighteen only when testing would offer immediate medical benefits or when there is a benefit to another family member and there is no anticipated harm to the person being tested. The committee considered genetic counseling before and after testing as essential components of the process.

The American Academy of Pediatrics Committee on Bioethics and Newborn Screening Task Force recommended the inclusion of tests in the newborn-screening battery based on scientific evidence. The academy advocated informed consent for newborn screening. (To date, most states do not require informed consent.) The Committee on Bioethics did not endorse carrier screening in people under eighteen years of age, except in the case of a pregnant teenager. It also recommended against predictive testing for adult-onset disorders in people under eighteen.

The American College of Medical Genetics, the American Society of Human Genetics (ASHG), and the World Health Organization have also weighed in about genetic testing of asymptomatic children, asserting that decision making should emphasize the child's well-being. One issue involves the value of testing asymptomatic children for genetic mutations associated with adult-onset conditions such as Huntington's disease. Because no treatment can begin until the onset of the disease, and at present there is no treatment to alter the course of the disease, it may be ill advised to test for it. Another concern is testing for carrier status of autosomal recessive or X-linked conditions such as cystic fibrosis or Duchenne muscular dystrophy. Experts caution that children might confuse carrier status with actually having the condition, which in turn might provoke needless anxiety.

There are, however, circumstances in which genetic testing of children may be appropriate and useful. Examples are children with symptoms of suspected hereditary disorders or those at risk for cancers in which inheritance plays a primary role.

ETHICAL CONSIDERATIONS—CHOICES AND CHALLENGES

Rapid advances in genetic research since the 1980s have challenged scientists, health care professionals, ethicists, government regulators, legislators, and consumers to stay abreast of new developments. Understanding the scientific advances and their implications is critical for everyone involved in making informed decisions about the ways in which genetic research and information will affect the lives of current and future generations. American citizens, scientists, ethicists, legislators, and regulators share this responsibility. According to the Human Genome Project Information Web site (February 5, 2003, http://genome.rtc.riken.go.jp/hgmis/elsi/elsi.html), the pivotal importance of these societal decisions was underscored by the allocation of 3% to 5% of the budget of the Human Genome Project for the study of ethical, legal, and social issues related to genetic research. To date, consideration of these issues has not produced simple or universally applicable answers to the many questions posed by the increasing availability of genetic information. Ongoing public discussion and debate is intended to inform, educate, and help people in every walk of life make personal decisions about their health and participate in decisions that concern others.

As researchers learn more about the genes responsible for a variety of illnesses, they can design more tests with increased accuracy and reliability to predict whether an individual is at risk of developing specific diseases. The ethical issues involved in genetic testing have turned out to be far more complicated than originally anticipated. Initially, physicians and researchers believed that a test to determine in advance who would develop or escape a disease would be welcomed by at-risk families, who would be able to plan more realistically about having children, choosing jobs, obtaining insurance, and going about their daily lives. Nevertheless, many people with family histories of a genetic disease have decided that not knowing is better than anticipating a grim future and an agonizing, slow death. They prefer to live with the hope that they will not develop the disease rather than having the certain knowledge that they will.

The discovery of genetic links and the development of tests to predict the likelihood or certainty of developing a disease raise ethical questions for people who carry a defective gene. Should women who are carriers of Huntington's disease or cystic fibrosis have children? Should a fetus with a defective gene be carried to term or aborted?

There are also concerns about privacy and the confidentiality of medical records and the results of genetic testing and possible stigmatization. Some people are reluctant to be tested because they fear they may lose their health, life, and disability insurances, or even their jobs if they are found to be at risk for a disease. Genetic tests are sometimes costly, and some insurers agree to reimburse for testing only if they are informed of the results. The insurance companies feel they cannot risk selling policies to people they know will become disabled or die prematurely.

The fear of discrimination by insurance companies or employers who learn the results of genetic testing is often justified. An insurance carrier may charge a healthy person a higher rate or disqualify an individual based on test results, and an employer might choose not to hire or to deny an affected individual a promotion. The American Society of Medical Genetics and most other medical professional associations agree that people should not be forced to choose between having a genetic test that could provide lifesaving information and avoiding a test to save a job or retain health insurance coverage.

Unanticipated Information and Results of Genetic Testing

Sometimes genetic testing yields unanticipated information, such as paternity, or other unexpected results, such as the presence of a disorder that was not sought directly. Many health professionals consider disclosure of such information, particularly when it does not influence health or medical treatment decision making, as counterproductive and even potentially harmful.

The ASHG is the primary professional organization for human geneticists in the United States. Its 8,000 members include researchers, academicians, clinicians, laboratory practice professionals, genetic counselors, nurses, and others involved in or with a special interest in human genetics. The ASHG recommends that family members not be informed of misattributed paternity unless the test requested was determination of paternity. The ASHG offers comparable guidelines regarding other unexpected results such as associations among diseases. For example, while performing screening for one disease, information about another disease may be discovered. Although the person may have requested screening for the first disorder, the presence of the second disorder may be unanticipated

and may lead to stigmatization and discrimination on the part of insurance companies and employers. The ASHG encourages all health professionals to educate, counsel, and obtain informed consents that include cautions regarding unexpected findings before performing genetic testing.

Advertising Genetic Testing Directly to Consumers

Although pharmaceutical drugs have been advertised directly to consumers for more than two decades, direct-to-consumer advertising of genetic tests is a relatively recent phenomenon. Especially controversial are direct-to-consumer DNA tests sold in retail stores and via the Internet that promise personalized nutritional counseling based on genetic makeup. In *Nutrigenetic Testing: Tests Purchased from Four Web Sites Mislead Consumers* (July 27, 2006, http://www.gao.gov/new .items/d06977t.pdf), Gregory Kutz of the U.S. Government Accountability Office asserts that these tests are of no medical value. He notes that by promising results they cannot deliver, they often deceive people. Although nutritional genomics and nutrigenetic testing, which consider how complex interactions between genes and diet may affect the risk of future illnesses, is a legitimate discipline, Kutz believes the firms offering these services make health claims that are not supported by credible scientific evidence. Kutz decries the current regulatory environment, which provides only limited oversight of firms developing and marketing new types of genetic tests.

FDA Moves to Regulate Genetic Testing

Experts have been urging either the FDA or the Centers for Medicare and Medicaid Services, which regulates clinical labs, to strengthen regulation of diagnostic genetic tests. Andrew Pollack reports in "F.D.A. Seeks to Regulate New Types of Diagnostic Tests" (*New York Times*, September 6, 2006) that the FDA took its first steps to regulate such tests by issuing in September 2006 a draft guideline that would require FDA approval before the test kits can be marketed. The FDA intends to regulate at least one category of genetic tests: those that measure multiple genes, proteins, or other pieces of clinical information taken from a patient and then analyze the data. The best known of such tests is Oncotype DX, which costs $3,500 and analyzes twenty-one genes in a sample of breast tumor and then computes a score that predicts whether the cancer will recur and whether the patient would benefit from chemotherapy.

Personal Choices and Psychological Consequences of Genetic Testing

The results of genetic tests may be used to make decisions such as whether to have children or end a pregnancy. Results that predict the likelihood that an individual will develop a disease may affect decisions about education, marriage, family, or career choices. The decision to undergo genetic testing and the results of the tests not only affect the individual tested but also his or her family members. For example, when an unaffected patient requests a genetic susceptibility test, another test of an affected relative may be required to accurately calculate probability. Family members may vary in their willingness to share genetic information and their desire to know about genetic risks.

There are psychological consequences of genetic testing and coming to terms with the results. People may be relieved or distressed when they learn the results of a genetic test. The results can change the way they feel about themselves and can influence their relationships with relatives. For example, family members who discover they are carriers for cystic fibrosis may feel isolated or estranged from siblings who have opted not to be tested. By contrast, those who find out they do not have a genetic mutation for Huntington's disease may feel guilty because they were spared and other family members were not, or they may worry about assuming responsibility for family members who develop the disease.

The complexity of genetic testing and the uncertainty of many results pose an additional psychological challenge. The results of predictive genetic tests are often expressed in probabilities rather than certainties, and, even for people with a high probability of developing a disease, there are often conflicting opinions about the most appropriate course of action. For example, a woman who tests positive for BRCA1 or BRCA2 has an increased lifetime risk of developing breast or ovarian cancer or both, but it does not mean that her risk of developing either or both is 100%. Depending on her personal circumstances and the medical advice she receives, she may opt to intensify screening to detect disease; use prescription medication intended to reduce risk, such as tamoxifen; or undergo preventive procedures, such as mastectomy and oophorectomy.

Even a negative test result can be stressful, creating nearly as many questions as it does answers about disease risk. A negative test result for BRCA1 or BRCA2 in a woman with an affected family member (one who has the genetic mutation) means that even though she does not have the genetic mutation, her risk is the same as that of the general population. There are multiple factors associated with the risk of developing breast cancer that are not identified through genetic testing, such as the age at which a woman has her first child. In addition, most breast cancer is not believed to be hereditary, and most women with diagnosed breast cancer do not test positive for BRCA1 or BRCA2 genetic mutations.

LEGISLATION TO PROTECT GENETIC INFORMATION AND PREVENT DISCRIMINATION

By 2007 most state legislatures had taken steps to safeguard genetic information beyond the protections provided for other types of health information. However, absent comprehensive federal legislation, not all people will be protected from discrimination based on genetic information.

Federal Executive Order 13145, "To Prohibit Discrimination in Federal Employment Based on Genetic Information," was signed on February 8, 2000. The executive order prohibits federal government agencies from obtaining genetic information from employees or job applicants and from using genetic information in hiring and promotion decisions. The executive order defines genetic information as information about an individual's genetic tests; information about the genetic tests of an individual's family members; and information about the occurrence of a disease, medical condition, or disorder in family members of the individual.

On October 14, 2003, the U.S. Senate passed the Genetic Information Nondiscrimination Act (S. 1053), which would have prohibited discrimination on the basis of genetic information in health insurance coverage and the workplace. Francis S. Collins, the director of the National Human Genome Research Institute, published a statement (October 14, 2003, http://genome.gov/11009127) expressing his contention that:

> No one should lose his job because of the genes he inherited. No one should be denied health insurance because of her DNA. But genetic discrimination affects more than jobs and insurance. It slows the pace of science. We know that many people have refused to participate in genetic research for fear of genetic discrimination. This means that without the kind of legal protections offered by this bill, our clinical research protocols will lack participants, and those who do participate will represent a self-selected group.

The Genetic Information Nondiscrimination Act of 2005 (S. 306) was unanimously passed on February 17, 2005. President George W. Bush expressed his support for the bill in a Statement of Administrative Policy. A virtually identical bill, the Genetic Information Nondiscrimination Act (H.R. 1227), was considered by the U.S. House of Representatives during 2005 but by the close of 2006 it had not made it out of committee.

Health Insurance Portability and Accountability Act of 1996

On August 21, 1996, President Bill Clinton signed the Health Insurance Portability and Accountability Act (HIPAA; PL 104-191). This legislation aims to provide better portability (transfer) of employer-sponsored insurance from one job to another. By preventing job lock—the need to remain in the same position or with the same employer for fear of losing health care coverage—the act was designed to afford American workers greater career mobility and the freedom to pursue job opportunities. Industry observers and policy makers viewed HIPAA as an important first step in the federal initiative to significantly reduce the number of uninsured people in the United States. They also hoped it would provide a measure of protection from genetics-based discrimination.

HIPAA stipulates that American workers who have previous insurance coverage are immediately eligible for new coverage when changing jobs. The law prohibits group health plans from denying new coverage based on past or present poor health and guarantees that employees can retain their health care coverage even after they leave their jobs. New employers can still require a routine waiting period (usually no more than three months) before paying for health benefits, but the new employee who applies for insurance coverage can be continuously covered during the waiting period. HIPPA also prohibits excluding an individual from group coverage because of past or present medical problems, including genetic information.

HIPAA does not, however, prohibit the use of genetic information as a basis for charging a group more for health insurance. It neither limits the collection of genetic information by insurers nor prohibits insurers from requiring an individual to take a genetic test. The act does not limit the disclosure of genetic information by insurers, and it does not apply to individual health insurers unless they are covered by the portability provision.

GENETIC COUNSELING

Generally, patients receive explicit counseling before undergoing genetic testing to ensure that they are able to make informed decisions about choosing to have the tests and the consequences of testing. Most genetic counseling is informative and nondirective—it is intended to offer enough information to allow families or individuals to determine the best courses of action for themselves but avoids making testing recommendations.

Patients undergoing tests to improve their care and treatment have different pretest counseling needs from those choosing susceptibility or predictive testing. In such instances genetic counselors do offer testing recommendations, particularly when a test offers an opportunity to prevent disease. As tests for genetic risk factors increasingly become routine in clinical medical practice, they are likely to be offered without formal pretest counseling. Genetic counselors urge physicians, nurses, and other health care professionals not to discount or rush through the process of obtaining informed consent to conduct a genetic test. They caution that potential psychological, social, and family implications should be acknowledged and addressed in advance of testing, including the potential for discrimination on the basis of genetic-risk status and the possibility that the predictive value of genetic information may be overestimated. The potentially life-changing consequences of genetic testing suggest that all health care professionals involved in the process should not only adhere to thoughtful informed consent procedures for genetic testing but also offer or make available genetic counseling when patients and families receive the results of testing.

CHAPTER 7

THE HUMAN GENOME PROJECT

[W]hen the full map of the human genome is known . . . we shall have passed through a phase of human civilisation as significant as, if not more significant than, that which distinguished the age of Galileo from that of Copernicus, or that of Einstein from that of Newton. . . . We have crossed a boundary of unprecedented importance. . . . There is no going back. . . . We are walking hopefully in the scientific foothills of a gigantic mountain range.

—Ian Lloyd, 1990

In 1953 James D. Watson and Francis H. C. Crick described the double helical structure of deoxyribonucleic acid (DNA). Their molecular DNA structure was published in *Nature* in April 1953, in an article that was little more than one page. Their article ushered in a new age of discovery in genetics and laid the foundation for the sequencing of the human genome.

The word *genome* was derived from two words: gene and chromosome. Today, *genome* is widely understood to be the entire complement of genetic material in the cell of an organism. A genome is composed of a series of four nitrogenous DNA bases: adenine (A), guanine (G), thymine (T), and cytosine (C). In each organism these bases are arranged in a specific order, or sequence, and this order constitutes the genetic code of the organism. In humans the genome is composed of approximately three billion bases. In 2001 a first draft sequence of the entire human genome was completed and made available to the public for study and research. The Human Genome Project (HGP) of the National Human Genome Research Institute (NHGRI), which is one of the National Institutes of Health (NIH), completed the full human genome sequence in April 2003.

LAYING THE GROUNDWORK FOR THE SEQUENCING OF THE HUMAN GENOME

During the 1960s and 1970s the techniques that would enable the study of molecular genetics were developed. In 1964 the American virologist Howard Temin worked with ribonucleic acid (RNA) viruses and discovered that Crick's central tenet—that DNA makes RNA, and RNA makes protein—did not always hold true. In 1965 Temin described the process of reverse transcriptase—that genetic information in the form of RNA could be copied into DNA. The enzyme called reverse transcriptase used RNA as a template for the synthesis of a complementary DNA strand. Throughout the 1960s the American biochemists Robert William Holley, Har Gobind Khorana, and Marshall Nirenberg, along with the American geneticist Philip Leder, all contributed to deciphering the genetic code by determining the DNA sequence for each of the twenty most common amino acids. Holley, Khorana, and Nirenberg were awarded the 1968 Nobel Prize in Physiology or Medicine.

The American biochemist Paul Berg created the first recombinant DNA in 1972, and his work paved the way for isolating and cloning genes. Recombinant DNA is formed by combining segments of DNA, frequently from different organisms. In 1975 the English molecular biologist Sir Edwin Southern developed a method to isolate and analyze fragments of DNA that remains in use today. Known as the Southern blot analysis, it is a technique for separating DNA fragments by electrophoresis (a technique that separates molecules based on their size and charge) and identifying a specific fragment using a DNA probe. Figure 7.1 shows how the Southern blot analysis is performed. It is used in genetic research, forensic examinations of DNA evidence in legal proceedings, and clinical medical practice.

In 1977 English biochemist Frederick Sanger, whose many accomplishments have been acknowledged by two Nobel Prizes, and his colleagues developed techniques to determine the nucleic acid base sequence for long sections of DNA. In 1978 American biologists Hamilton O. Smith and Daniel Nathans and the Swiss molecular biologist Werner Arber were awarded the Nobel Prize for an

FIGURE 7.1

Southern blot analysis

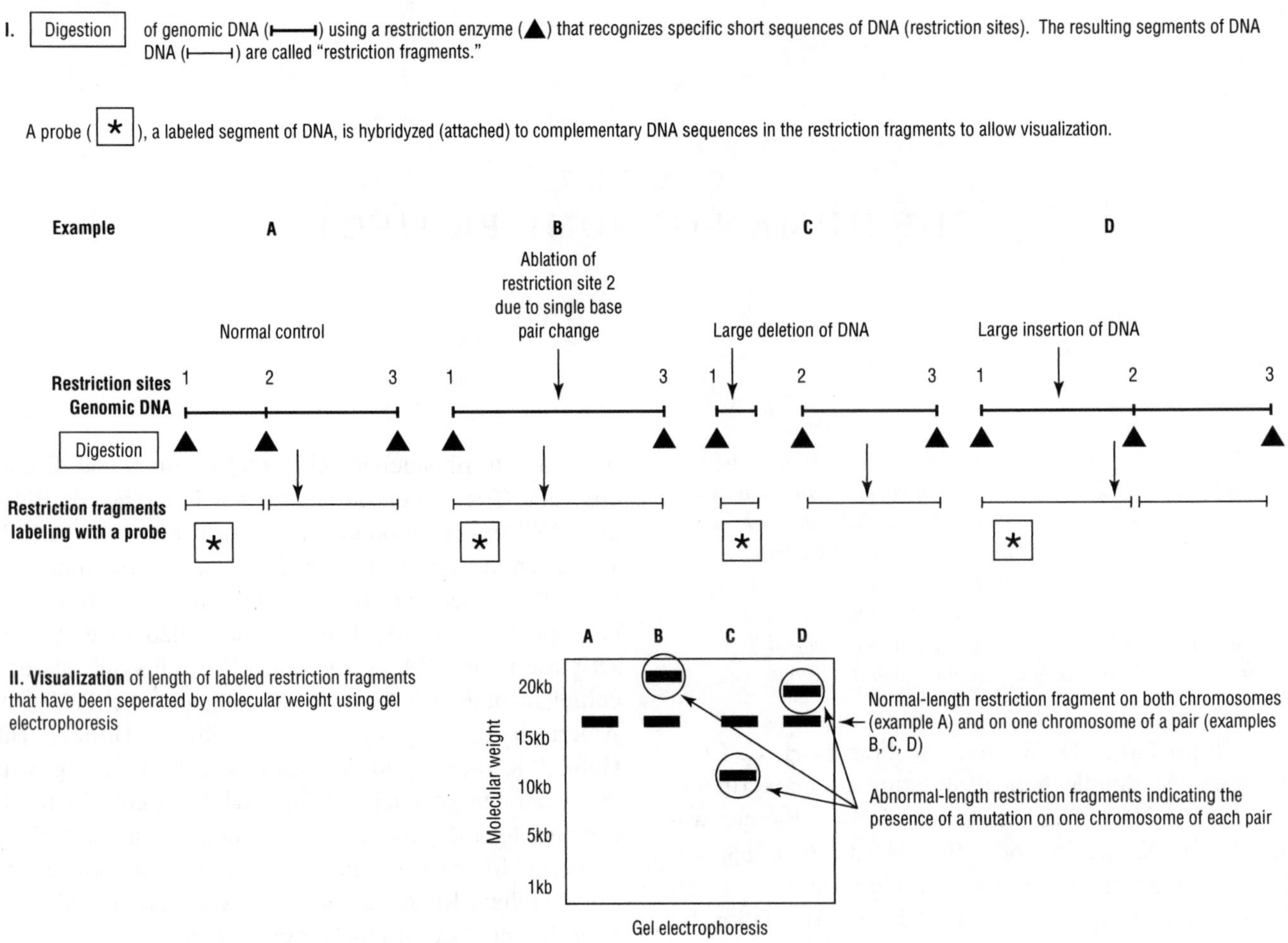

I. [Digestion] of genomic DNA (├────┤) using a restriction enzyme (▲) that recognizes specific short sequences of DNA (restriction sites). The resulting segments of DNA DNA (├────┤) are called "restriction fragments."

A probe ([★]), a labeled segment of DNA, is hybridyzed (attached) to complementary DNA sequences in the restriction fragments to allow visualization.

Example

A — Normal control

B — Ablation of restriction site 2 due to single base pair change

C — Large deletion of DNA

D — Large insertion of DNA

Restriction sites / Genomic DNA

[Digestion]

Restriction fragments labeling with a probe

II. **Visualization** of length of labeled restriction fragments that have been seperated by molecular weight using gel electrophoresis

Molecular weight: 20kb, 15kb, 10kb, 5kb, 1kb

Gel electrophoresis

Normal-length restriction fragment on both chromosomes (example A) and on one chromosome of a pair (examples B, C, D)

Abnormal-length restriction fragments indicating the presence of a mutation on one chromosome of each pair

SOURCE: "Southern Blot," in *GeneTests Data Base: GeneReviews Illustrated Glossary*, National Institutes of Health, U.S. National Library of Medicine, National Human Genome Research Institute, http://www.genetests.org/servlet/access?qry=ALLTERMS&db=genestar&fcn=term&gtreport2=true&id=8888891&key=mbmumbAd9rOWs (accessed October 26, 2006)

---

array of discoveries made during the 1960s, including the use of restriction enzymes, which ignited the biotechnology field. Restriction enzymes recognize and cut specific DNA sequences. The same year restriction fragment length polymorphisms (DNA sequence variants) were discovered. Figure 7.2 shows a single nucleotide polymorphism—single base changes between homologous DNA fragments.

Using these new genetic techniques, several genes for serious human disorders were identified during the 1980s. In 1982 the American molecular biologist James F. Gusella and his colleagues at Harvard University began studying patients with Huntington's disease and determined that the gene for this degenerative, neuropsychiatric disorder was located on the short arm of chromosome 4. That same year a gene for neurofibromatosis type I was found on the long arm of chromosome

17. Neurofibromatoses are a group of genetic disorders that cause tumors to grow along various types of nerves and can affect the development of nonnervous tissues such as bones and skin. The disorder may also result in developmental abnormalities such as learning disabilities.

In 1985 the American biochemist Kary B. Mullis and his colleagues at the Cetus Corporation in California pioneered the polymerase chain reaction, a fast, inexpensive technique that amplifies small fragments of DNA to make sufficient quantities available for DNA sequence analysis—that is, determining the exact order of the base pairs in a segment of DNA. Because it enabled researchers to make an unlimited number of copies of any piece of DNA, it was dubbed "molecular photocopying," and in 1993 Mullis was awarded the Nobel Prize for this tremendous breakthrough in gene analysis. By 1987 automated sequencers were developed, enabling even more

FIGURE 7.2

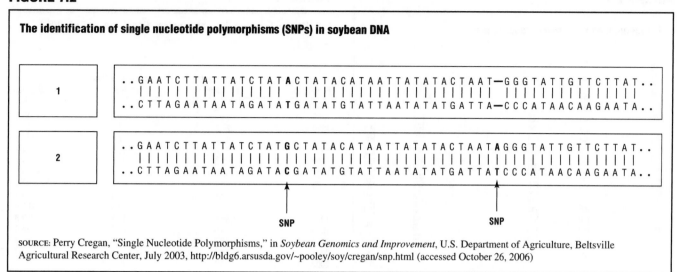

The identification of single nucleotide polymorphisms (SNPs) in soybean DNA

SOURCE: Perry Cregan, "Single Nucleotide Polymorphisms," in *Soybean Genomics and Improvement*, U.S. Department of Agriculture, Beltsville Agricultural Research Center, July 2003, http://bldg6.arsusda.gov/~pooley/soy/cregan/snp.html (accessed October 26, 2006)

FIGURE 7.3

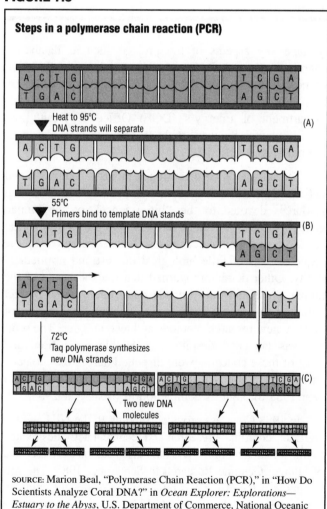

Steps in a polymerase chain reaction (PCR)

SOURCE: Marion Beal, "Polymerase Chain Reaction (PCR)," in "How Do Scientists Analyze Coral DNA?" in *Ocean Explorer: Explorations—Estuary to the Abyss*, U.S. Department of Commerce, National Oceanic and Atmospheric Administration, July 2005, http://oceanexplorer.noaa.gov/explorations/04etta/background/dna/dna.html (accessed December 14, 2006)

rapid sequencing and analysis on large segments of DNA. Figure 7.3 shows the steps involved in a polymerase chain reaction.

In 1985 the Canadian molecular geneticist Lap-Chee Tsui and his research team mapped the gene responsible for cystic fibrosis, the most common inherited fatal disease of children and young adults in the United States, to the long arm of chromosome 7. The gene for cystic fibrosis was discovered in 1989, and it was determined that three missing nucleic acid bases occurred in the altered gene of 70% of patients with cystic fibrosis.

The mutations associated with Duchenne muscular dystrophy were identified in 1987. This gene is located close to the gene for chronic granulomatous disease (an X-linked autosomal recessive disorder that, if left untreated, is fatal in childhood) on the short arm of the X chromosome. In 1990 the American geneticist Mary-Claire King found the first evidence that a gene on chromosome 17 (now known as BRCA1) could potentially be associated with an inherited predisposition to breast and ovarian cancer.

The discoveries and technological advances made by researchers during the 1970s and 1980s gave rise to modern clinical molecular genetics. The study of chromosome structure and function, called cytogenetics, produced methods to view distinct bands on each chromosome. Figure 7.4 is a cytogenetic map of human chromosomes. Cytogenetic studies are applied in three broad areas of medicine: congenital (from birth) disorders, prenatal diagnosis, and neoplastic diseases (cancer).

BIRTH OF THE HUMAN GENOME PROJECT

The first meetings to discuss the feasibility of sequencing the human genome were organized by Robert Louis Sinsheimer, a molecular biologist and chancellor of

FIGURE 7.4

Cytogenic map of human chromosomes

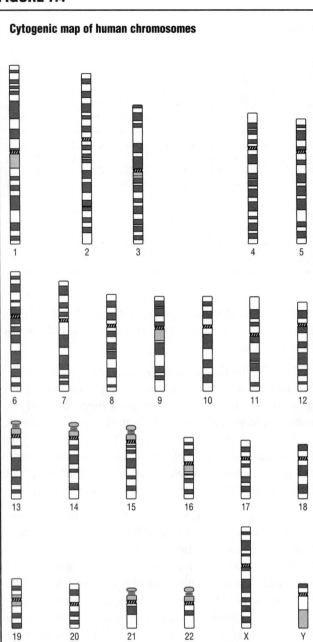

SOURCE: "Cytogenic Map," in *Talking Glossary of Genetic Terms*, U.S. Department of Health and Human Services, National Institutes of Health, National Human Genome Research Institute, Division of Intramural Research, http://www.genome.gov/Pages/Hyperion//DIR/VIP/Glossary/Illustration/cytogenetic_map.shtml (accessed October 26, 2006)

TABLE 7.1

U.S. Human Genome Project funding, fiscal years 1988–2003

[$ millions]

Fiscal year	Department of Energy	National Institutes of Health	U.S. total
1988	10.7	17.2	27.9
1989	18.5	28.2	46.7
1990	27.2	59.5	86.7
1991	47.4	87.4	134.8
1992	59.4	104.8	164.2
1993	63.0	106.1	169.1
1994	63.3	127.0	190.3
1995	68.7	153.8	222.5
1996	73.9	169.3	243.2
1997	77.9	188.9	266.8
1998	85.5	218.3	303.8
1999	89.9	225.7	315.6
2000	88.9	271.7	360.6
2001	86.4	308.4	394.8
2002	90.1	346.7	434.3
2003	64.2	372.8	437

Note: These numbers do not include construction funds, which are a very small part of the budget.

SOURCE: "Human Genome Project Budget," in *Human Genome Project Information*, U.S. Department of Energy Office of Science, Office of Biological and Environmental Research, Human Genome Project, http://www.ornl.gov/sci/techresources/Human_Genome/project/budget. shtml (accessed October 27, 2006)

manage key aspects of the project, such as databases, sharing of research findings and materials, and cultivation of new technologies. The initial funding from Congress was $17.3 million to the NIH and $11.8 million to the U.S. Department of Energy's (DOE) Office of Health and Environmental Science, with progressive increases over the next few years.

The allocation of these funds incited an impassioned debate. Opponents argued that the financial and human resources devoted to the "big science" of the human genome project would divert research funds from vital scientific and biomedical research and that most of the sequence was of little biological interest and no medical utility. Other detractors warned that the sheer size of the human genome would impede completion of the project within a reasonable time frame without the creation of entirely new research methods and technologies. The project was launched despite considerable opposition, and most of these concerns were dispelled during the project's early years. Table 7.1 shows the HGP budget from fiscal year 1988 through its completion in fiscal year 2003.

In 1988 Congress provided funding to the NIH and the DOE to "coordinate research and technical activities related to the human genome." The NIH also established the Office of Human Genome Research in September 1988. The following year the office was renamed the National Center for Human Genome Research (NCHGR). James Watson served as its enthusiastic champion and director until April 1992. Following his appointment, the NIH and the DOE committed 3% to 5% of the project's budget to address ethical,

the University of California (UC) at Santa Cruz, and were held on campus in 1985. The idea of sequencing the human genome generated excitement among the many well-known researchers in attendance—they considered the undertaking to be the "Holy Grail" of molecular biology. The following year the U.S. Congress began to consider the feasibility of human genome research. Congress did not, however, decide to fund the project until 1988, after it concluded that the establishment of administrative centers accountable to Congress could effectively legal, and social issues that arose from the study of the human genome (February 5, 2003, http://genome.rtc.riken.go.jp/hgmis/elsi/elsi.html). This ambitious undertaking constituted the largest bioethics program, in terms of funding and human resources, in the world.

The Human Genome Project Information Web site (December 7, 2005, http://www.ornl.gov/sci/techresources/Human_Genome/project/about.shtml) describes the ambitious goals of the HGP when it began in 1990. The overarching HGP goals were to:

- Identify all the approximately 30,000 genes in human DNA

- Determine the sequences of the three billion chemical base pairs in human DNA

- Store HGP findings and other information in databases

- Improve tools for data analysis

- Transfer related technologies to the private sector

- Effectively address the ethical, legal, and social issues that might arise from the project

International genomic research was also under way in England, France, Germany, Japan, and other countries. In 1987 the Italian National Research Council launched a genome research project; the United Kingdom began its project in February 1989. In 1989 an international group of geneticists founded the Human Genome Organization (HUGO) in Switzerland. Many international collaborations had already been forged as individual scientists exchanged information in their quests for genetic links to disease. HUGO developed an international framework to coordinate research projects and prevent wasted resources through duplication, creating a culture of sharing data. In 1990 the European Commission initiated a two-year human genome project. Russia funded its genome research project in the same year.

In April 1990 the initial planning stage of the U.S. HGP was completed with the publication of the joint research plan *Understanding Our Genetic Inheritance: The Human Genome Project—The First Five Years, Fiscal Years 1991–1995* (http://www.ornl.gov/sci/techresources/Human_Genome/project/5yrplan/summary.shtml). Just two years into the five-year plan, Watson resigned from his leadership position with the NCHGR because he vehemently disagreed with NIH decisions about the commercialization, propriety, and legality of patenting human gene sequences. Watson maintained that data from the HGP should be in the public domain and freely available to all scientists as well as to the public. In April 1993 the American geneticist and physician Francis S. Collins was named the director.

Many prominent researchers sided with Watson against the patenting and commercialization of HGP data.

TABLE 7.2

Model organisms sequenced

Date sequenced[a]	Species	Total bases[b]
7/28/1995	*Haemophilus influenzae* (bacterium)	1,830,138
10/30/1995	*Mycoplasma genitalium* (bacterium)	580,073
5/29/1997	*Saccharomyces cerevisiae* (yeast)	12,069,247
9/5/1997	*Escherichia coli* (bacterium)	4,639,221
11/20/1997	*Bacillus subtillis* (bacterium)	4,214,814
12/31/1998	*Caenorhabditis elegans* (round worm)	97,283,371
		99,167,964[c]
3/24/2000	*Drosophila melanogaster* (fruit fly)	~137,000,000
12/14/2000	*Arabidopsis thaliana* (mustard plant)	~115,400,000
1/26/2001	*Oryza sativa* (rice)	~430,000,000
2/15/2001	*Homo sapiens* (human)	~3,200,000,000

[a]First publication date.
[b]Data excludes organelles or plasmids. These numbers should not be taken as absolute. Scientists are confirming the sequences; several laboratories were involved in the sequencing of a particular organism and have slightly different numbers; and there are some strain variations.
[c]The first number was originally published, and the second is a correction as of June 2000.

SOURCE: Richard Robinson, ed., "Model Organisms Sequenced," in *Genetics*, Vol. 2, *E–I*, Macmillan Reference USA, 2002

In 1996 scientists at leading research institutions throughout the world agreed to submit their findings and genome sequences to GenBank, a genome database maintained by the NIH. In a resounding and unanimous move, they required the publication of any submitted sequence data on the Internet within twenty-four hours of its receipt by GenBank. This action ensured that gene sequences were in the public domain and could not be patented.

Worming Away

Although sequencing the human genome was the principal objective, the HGP also sought to sequence the genomes of other organisms. These other organisms served as models, enabling researchers to test and refine new methods and technologies that helped identify corresponding genes in the human genome. Table 7.2 is a list of some of the model organisms, including the roundworm, sequenced during the course of the HGP, along with the dates the sequences were published and the number of bases in each.

At England's Cambridge University, the geneticist and molecular biologist Sydney Brenner was studying the nematode worm *Caenorhabditis elegans*. By 1989 Brenner and his colleagues had successfully produced a map of the entire *Caenorhabditis elegans* genome. The map consisted of multiple overlapping fragments of DNA, arranged in the correct order, and Brenner's research team printed the worm's genome on postcard-sized pieces of paper.

Watson believed that the genomes of smaller organisms would not only help to refine research methods and the technology but also provide valuable sources of comparison once the human genome project was under way. The worm map convinced Watson that *Caenorhabditis elegans* should be the first multicellular organism to have its complete genome accurately sequenced. When the worm-sequencing project began in 1990, the first automatic sequencing machines had just become available from Applied Biosystems, Inc. The sequencing machines enabled the worm pilot project to meet its objective of sequencing three million bases in three years. Equally important, the worm project demonstrated that the technology could scale up—that is, more machines and more technologists could produce more sequences faster.

A DECADE OF ACCOMPLISHMENTS: 1993 TO 2003

When the HGP began in September 1990, its projected completion date was 2005, at a cost of $3 billion for just the U.S. portion of the research. Ever-improving research techniques—including the use of restriction fragment length polymorphisms, which is described in detail in Figure 7.5, as well as the polymerase chain reaction, bacterial and yeast artificial chromosomes, and pulsed-field gel electrophoresis—accelerated the progress of the project. The HGP researchers finished the mapping two years earlier than scheduled, and U.S. scientists spent just $2.7 billion.

Although the HGP was truly a collaborative, international effort, most of the sequencing work was performed at the Whitehead Institute for Medical Research in Massachusetts, the Baylor College of Medicine in Texas, the University of Washington, the Joint Genome Institute in California, and the Sanger Centre in the United Kingdom. Along with U.S government funding, the HGP was supported by the Wellcome Trust, a charitable foundation in the United Kingdom.

Public and Private Initiatives Compete

In 1993 the NCHGR established a Division of Intramural Research, charged with developing genome technology research of specific diseases. By 1996 eight NIH institutes and centers had also collaborated to create the Center for Inherited Disease Research to study the genetics of complex diseases. In 1997 the NCHGR gained full institute status at the NIH, becoming the National Human Genome Research Institute (NHGRI). A third five-year plan was announced in 1998 in *Science*.

The mid-1990s also saw the birth of a privately funded genomics effort led by the American geneticist J. Craig Venter. Venter had been working in a laboratory at the NIH when he decided to concentrate his sequencing efforts not on the genome itself, but on the gene products—that is, messenger RNAs produced by each cell. Besides the genes themselves, the genome is composed of a much larger amount of DNA whose function is as yet unknown. This portion of the genome is often referred to as "junk DNA," although it is possible that its true value has not yet been determined. Venter eventually left the NIH to establish a private, nonprofit organization, The Institute for Genomic

Research (TIGR), an organization aimed at collecting and interpreting vast numbers of expressed sequence tags. (Expressed sequence tags are random fragments of complementary DNAs derived from the information in the RNAs, which contain all the information that is actually expressed in a given cell type.) In July 1995 TIGR published the first sequenced genome of the bacteria *Haemophilus influenzae*. (See Table 7.2.)

In May 1998 Venter announced that he was allying with the Perkin-Elmer Corporation to form a new company that would compete directly with the public effort to sequence the entire human genome by 2001. The new company would become Celera Genomics. Venter planned to take a different approach from the one used by the HGP. Instead of the map-based, gene-by-gene approach taken by the HGP, his firm planned to break the genome into random lengths, and then sequence and reassemble it. This method saved time by eliminating the mapping phase, but it required robust computing capabilities to reassemble the human genome, which includes many repeated sequences. Venter's approach relied on the use of a supercomputer and 300 high-speed automatic sequencers manufactured by the Perkin-Elmer Corporation; it was the precursor to the large-scale genomics studies that have become standard.

The competition between public and private initiatives began in earnest. Within one week of the launch of the private initiative, the Wellcome Trust increased its funding to the Sanger Centre to step up the production of raw sequence. In response to this increased support, the Sanger Centre revised its objective by aiming to sequence a full one-third of the entire genome rather than just one-sixth. The race between the public and private projects was on, and the milestones of the sequencing project came ever more rapidly.

In the United States Venter's firm energized the publicly funded project and inspired intensified efforts. HGP investigators feared that if their efforts were perceived as slow and inefficient, the HGP would lose congressional support and funding. The threat of genomic information ending up solely in the hands of a private firm was simply unacceptable to the researchers. When Celera entered the race to sequence the human genome, it changed the landscape of the field. Celera resolved to make its data available only to paying customers and planned to patent some sequences before releasing them.

The publicly funded HGP released sequence information quickly both to provide the scientific community with timely data that were immediately usable and to place identified sequences beyond the reach of commercial companies wishing to patent them or charge for access to the data. The leadership of the HGP echoed the sentiments that had prompted Watson's resignation. They contended that patenting the human sequence was unethical and delayed the timely application of genomic information to medical disorders.

FIGURE 7.5

Restriction fragment length polymorphism analysis (RFLP analysis)

Step 1: Digestion. Amplicons (PCR-amplified DNA segments) are cut (▲) with a restriction enzyme wherever a specific DNA sequence occurs.

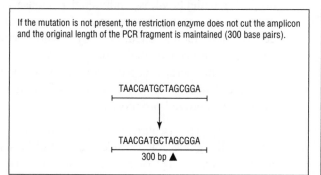

If the mutation is not present, the restriction enzyme does not cut the amplicon and the original length of the PCR fragment is maintained (300 base pairs).

TAACGATGCTAGCGGA
↓
TAACGATGCTAGCGGA
300 bp ▲

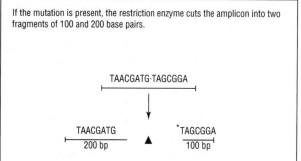

If the mutation is present, the restriction enzyme cuts the amplicon into two fragments of 100 and 200 base pairs.

TAACGATG·TAGCGGA
↓
TAACGATG   *TAGCGGA
200 bp   ▲   100 bp

Three outcomes are possible:

Patient A	Patient B	Patient C
300 bp ▲	200 bp ▲ 100 bp	200 bp ▲ 100 bp
300 bp ▲	300 bp ▲	200 bp ▲ 100 bp
Restriction enzyme does not cut amplicons from either allele	Restriction enzyme cuts amplicons from one allele	Restriction enzyme cuts amplicons from both alleles

Step 2: Visualization of cut and uncut amplicons on an electrophoretic gel.
The size of the DNA fragment determines the distance it migrates on the gel; short fragments travel farther than long fragments.

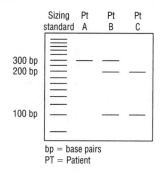

Sizing standard | Pt A | Pt B | Pt C

300 bp
200 bp

100 bp

bp = base pairs
PT = Patient

Step 3: Interpretation.

Patient A: No copies of mutation
Patient B: Heterozygous for mutation
Patient C: Homozygous for mutation

SOURCE: "Restriction Fragment Length Polymorphism Analysis (RFLP Analysis)," in *GeneTests Data Base: GeneReviews Illustrated Glossary*, National Institutes of Health, U.S. National Library of Medicine, National Human Genome Research Institute, http://www.genetests.org/servlet/access?qry=ALLTERMS&db=genestar&fcn=term&gtreport2=true&id=8888891&key=mbmumbAd9rOWs (accessed October 27, 2006)

The international partners in the genome project met in Bermuda in February 1996 at a strategy meeting sponsored by the Wellcome Trust. There they created the "Bermuda Principles," a set of conditions that govern access to data, including the standard that sequence information be released into public databases within twenty-four hours. To adhere to this agreement, participating scientists were to deposit base sequences into one of three databases within twenty-four hours of sequencing completion. The data contained in the three databases were exchanged daily. Because these were public databases, access to the stored sequences was free and unrestricted. The agreement was extended to data on other organisms at a meeting later that same year.

FIRST DRAFT OF THE COMPLETED HUMAN GENOME

In April 2000 Celera announced that it was prepared to present the first draft of the human genome. Not unexpectedly, scientists and the public eagerly anticipated this "first look" at the human genome. Although geneticists and other scientists could better comprehend the mechanics and the future implications of this endeavor than the general public, the significance of this achievement was evident to professionals and laypeople alike. The professional literature and the mass media had successfully communicated the importance of this achievement, and it was understood that knowledge of the human genome held the key to the singularity of the human species. Furthermore, it was widely assumed that this information would be the basis for unprecedented advances in medicine and biomedical technology.

In February 2001 the first working draft of the human genome—90% of the sequence of the genome's three billion bases—was published in special issues of the journals *Nature* (February 15, 2001) and *Science* (February 16, 2001). *Nature* detailed initial analysis of the descriptions of the sequence generated by the public HGP, and *Science* contained the draft sequence reported by private projects conducted by Celera.

One of several surprises from the first draft was that previous estimates of gene number appeared to have been wildly inaccurate. Most pregenome project estimates predicted that humans had as many as 60,000 to 150,000 genes. The first draft of the complete genome sequence indicated that the true number of genes required to make a human being was less than 40,000. By comparison, yeast have about 6,000 genes, fruit flies have 14,000, roundworms have 19,000, and the mustard weed plant has 26,000. Another surprise was the observation that humans share 99.9% of the nucleotide code in the human genome. Notably, human diversity at the genetic level is encoded by less than a 0.1% variation in DNA.

First Draft Is Headline News

A White House press release (June 25, 2000, http://www.ornl.gov/sci/techresources/Human_Genome/project/clinton1.shtml) predicted some of the anticipated medical outcomes of the project. These included the ability to:

- Alert patients that they are at risk for certain diseases. Once scientists discover which DNA sequence changes in a gene can cause disease, healthy people can be tested to see whether they risk developing conditions such as diabetes or prostate cancer later in life. In many cases, this advance warning can be a cue to start a vigilant screening program, to take preventive medicines, or to make diet or lifestyle changes that may prevent the disease.

- Reliably predict the course of disease. Diagnosing ailments more precisely will lead to more reliable predictions about the course of a disease. For example, a genetic fingerprint will allow doctors treating prostate cancer to predict how aggressive a tumor will be. New genetic information will help patients and doctors weigh the risks and benefits of different treatments.

- Precisely diagnose disease and ensure that the most effective treatment is used. Genetic analysis allows us to classify diseases, such as colon cancer and skin cancer, into more defined categories. These improved classifications will eventually allow scientists to tailor drugs for patients whose individual response can be predicted by genetic fingerprinting. For example, cancer patients facing chemotherapy could receive a genetic fingerprint of their tumor that would predict which chemotherapy choices are most likely to be effective, leading to fewer side effects from the treatment and improved prognoses.

- Developing new treatments at the molecular level. Drug design guided by an understanding of how genes work and knowledge of exactly what happens at the molecular level to cause disease will lead to more effective therapies. In many cases, rather than trying to replace a gene, it may be more effective and simpler to replace a defective gene's protein product. Alternatively, it may be possible to administer a small molecule that would interact with the protein to change its behavior. This is the strategy behind a drug in development for chronic myelogenous leukemia, which targets the genetic flaw causing the disease. It attaches to the abnormal protein caused by the genetic flaw and blocks its activity. In preliminary tests, blood counts returned to normal in all patients treated with the drug.

On June 26, 2000, *BBC News* compiled assessments of the achievement by the world's premier scientists and politicians (http://news.bbc.co.uk/1/hi/sci/tech/807126.stm). Venter spoke for many researchers when he said, "I think we will view this period as a very historic time, a new starting point." Michael Dexter of the Wellcome Trust echoed Venter's sentiments when he ventured, "This is the outstanding achievement not only of our lifetime, but in terms of human history. I say this, because the Human Genome Project does have the potential to impact on the life of every person on this planet." Randy Scott, the president of Incyte, another private firm involved in genomics research, predicted that "[t]he availability of genome sequence is just the beginning. Scientists now want to understand the genes and the role they play in the prevention, diagnosis and treatment of disease." Mike Stratton of the Cancer Genome Project was equally optimistic when he said that "[i]t would surprise me enormously if in twenty years the treatment of cancer had not been transformed." The Nobel Prize–winning English biochemist and professor Frederick Sanger, the pioneer of DNA sequencing, expressed the collective awe of the scientific community when the HGP was completed earlier than anticipated. Sanger admitted, "I never thought it would be done as quickly as this."

The U.S. media celebrated the achievement with a flood of press releases and features. Efforts were also

FIGURE 7.6

From DNA to humans

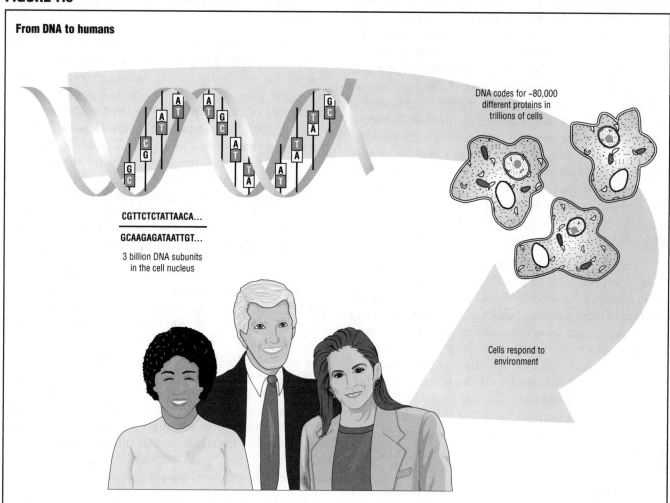

CGTTCTCTATTAACA...

GCAAGAGATAATTGT...

3 billion DNA subunits
in the cell nucleus

DNA codes for ~80,000
different proteins in
trillions of cells

Cells respond to
environment

SOURCE: "From DNA to Humans," in "Image Gallery," in *Human Genome Project Information*, U.S. Department of Energy Office of Science, Office of Biological and Environmental Research, Human Genome Project, http://www.ornl.gov/sci/techresources/Human_Genome/graphics/slides/images/00-0482 .jpg (accessed October 27, 2006)

TABLE 7.3

Projected national benefits of genomics research, 2020, 2040, and 2050

Within a decade		Long term
Develop knowledge base for cost-effective cleanup strategies	→	**2020** Save billions of dollars in toxic waste cleanup and disposal
Understand earth's natural carbon cycle and design strategies for enhanced carbon capture	→	**2040** Help stabilize atmospheric carbon dioxide to counter global warming
Increase biological sources of fuels and electricity	→	**2050** Contribute to U.S. energy security • Biohydrogen-based industry in place

SOURCE: "Payoffs for the Nation," in "Image Gallery," in *Human Genome Project Information*, U.S. Department of Energy Office of Science, Office of Biological and Environmental Research, Human Genome Project, http:// www.ornl.gov/sci/techresources/Human_Genome/graphics/slides/images/ payoffs.jpg (accessed October 27, 2006)

made to explain this monumental accomplishment to the public and to educate students. The DOE Human Genome Program provided a wealth of information about the HGP and its findings on the Internet. Figure 7.6 is an example of the information the DOE made available to the public. Table 7.3 presents some of the potential environmental benefits of the HGP that may be realized in the future.

PUFFER FISH AND MOUSE GENOMES ARE SEQUENCED

In July 2002 the DOE Joint Genome Institute (JGI), operated by the Lawrence Berkeley National Laboratory, the Lawrence Livermore National Laboratory, and the Los Alamos National Laboratory, announced the draft sequencing, assembly, and analysis of the genome of the Japanese puffer fish *Fugu rubripes*. The Fugu Genome Project was initiated in 1989 in Cambridge, England, and in November 2000 the International Fugu

Genome Consortium was formed, headed by the JGI. During 2001 the puffer fish genome was sequenced and assembled using the whole genome method pioneered by Celera. The puffer fish was the first vertebrate genome to be publicly sequenced and assembled in this manner and the first vertebrate genome published after the human genome. According to the JGI (2005, http://genome .jgi-psf.org/Takru4/Takru4.home.html), puffer fish have the smallest known genomes among vertebrates (animals with bony backbones or cartilaginous spinal columns—fish, reptiles, birds, and mammals, including humans). The puffer fish sequence has about the same number of genes as the considerably larger human genome, but is more compact because it contains relatively little of the junk DNA present in the human genome sequence.

Comparison of the human and puffer fish genomes enabled investigators to predict the existence of nearly 1,000 previously unidentified human genes. Although the function of these additional genes is as yet unknown, they contribute to the complete catalog of human genes. Ascertaining the existence and location of genes helped scientists begin to describe how they are regulated and function in the human body. Of the more than 30,000 puffer fish genes identified, the vast majority of human genes have counterparts in the puffer fish, with the most significant differences in genes of the immune system, metabolic regulation, and other physiological systems that are not alike in fish and mammals.

On December 5, 2002, the first draft of the sequence of the mouse genome was published in *Nature*. The mouse genome findings were deemed among the most important in terms of their comparability with humans. Mice and humans have about the same number of genes—approximately 20,000—and DNA base pairs—mice have 2.5 billion and humans have 2.9 billion. More important, about 90% of genes associated with medical disorders in humans have counterparts in mice. This finding means that mice are especially well suited for studying diseases that afflict humans and for testing therapeutic treatments for disease.

HUMAN GENOME PROJECT IS COMPLETED

After the publication of a first-draft human genome in 2001, work continued to fill in the blanks and produce a complete and accurate sequence. On January 10, 2003, another milestone in the human genome sequencing effort was reported: the fourth human chromosome—chromosome 14, the largest one to date, with 87 million base pairs—had been sequenced. The researchers Jean Weissenbach and Roland Heilig of Genoscope, France's National Sequencing Center in Paris, published their findings online in *Nature*. They found two genes that are vital for immune responses on chromosome 14 and about sixty genes that, when defective, contribute

to disorders such as spastic paraplegia and Alzheimer's disease.

In a remarkable coincidence that made the crowning achievement of the HGP even more poignant, the completion of the sequencing of the human genome occurred during the same year slated for celebrations of the fiftieth anniversary of the discovery of the DNA double helix. On April 14, 2003, the International Human Genome Sequencing Consortium, directed in the United States by the DOE and the NHGRI, announced the successful completion of the HGP more than two years earlier than had been anticipated (http://www.genome.gov/11006929). The International Human Genome Sequencing Consortium included scientists at twenty sequencing centers in China, France, Germany, Great Britain, Japan, and the United States.

Nature, the same journal that had published the groundbreaking discoveries of Watson and Crick fifty years earlier, hailed the era of the genome in a special edition dated April 24, 2003. As had been the practice since the inception of the HGP, the entirety of sequence data generated by the HGP was immediately entered into public databases and made freely available to the scientific community throughout the world, with no restrictions on its use or redistribution. The data are used by researchers in academic settings and industry, as well as by commercial firms that provide information services to biotechnologists. Figure 7.7 shows some of the landmarks in the sequenced human genome.

In the April 14, 2003, press release "All Goals Achieved: New Vision for Genome Research Unveiled" (http://www.genome.gov/11006929), the NHGRI described the international effort to sequence the three billion DNA base pairs in the human genome as "one of the most ambitious scientific undertakings of all time," comparing it to feats such as splitting the atom or traveling to the moon. NHGRI director Francis S. Collins proudly declared that "the Human Genome Project has been an amazing adventure into ourselves, to understand our own DNA instruction book, the shared inheritance of all humankind. All of the project's goals have been completed successfully—well in advance of the original deadline and for a cost substantially less than the original estimates." In the same press release Eric Lander, the director of the Whitehead Institute/Massachusetts Institute of Technology Center for Genome Research, predicted the postgenomic era when he asserted, "The Human Genome Project represents one of the remarkable achievements in the history of science. Its culmination this month signals the beginning of a new era in biomedical research. Biology is being transformed into an information science, able to take comprehensive global views of biological systems. With knowledge of all the components of the cells, we will be able to tackle biological problems at their most fundamental level."

FIGURE 7.7

Selected landmarks of the Human Genome Project

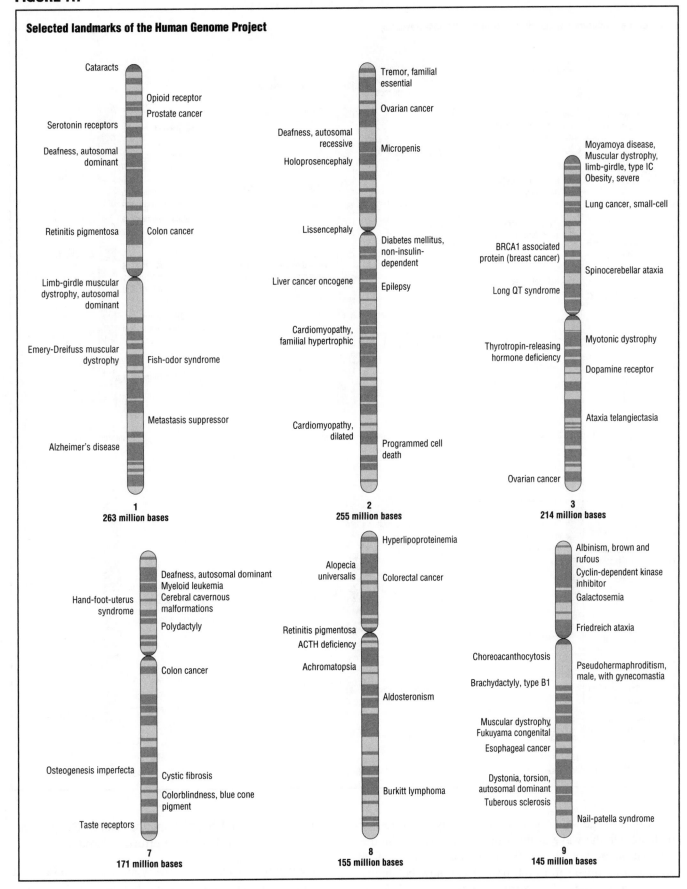

FIGURE 7.7

Selected landmarks of the Human Genome Project [CONTINUED]

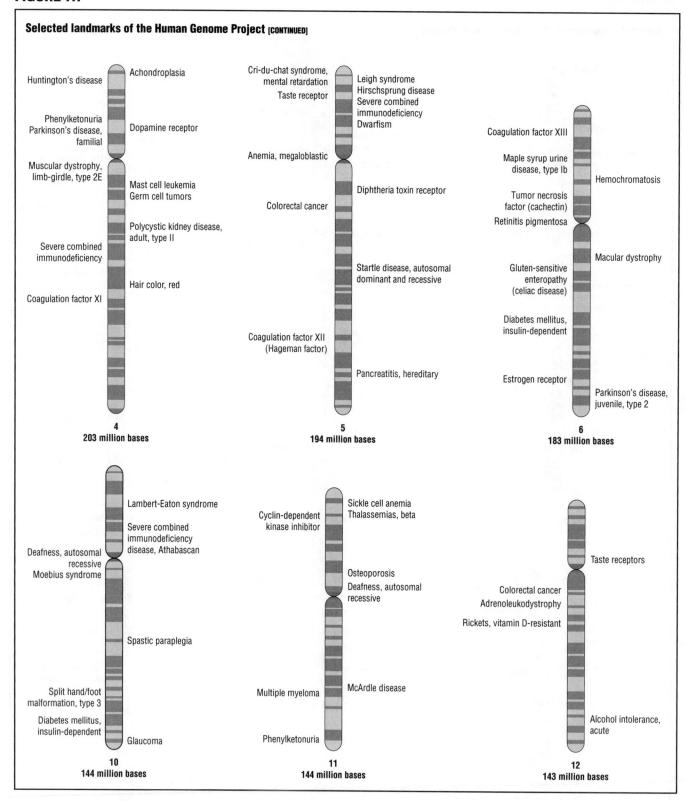

Collins urged the scientific community not to rest on its laurels in the wake of this triumph, saying, "With this foundation of knowledge firmly in place, the medical advances promised from the project can now be significantly accelerated." The April 24, 2003, issue of *Nature* detailed the challenges researchers would face in the postgenomic era as they sought to employ the HGP data to treat disease and improve public health. Recommendations included collaborative efforts to produce:

- New tools to allow discovery in the not-too-distant future of the genetic contributions to frequently occurring diseases, including diabetes, heart disease, and mental illnesses such as schizophrenia

FIGURE 7.7

Selected landmarks of the Human Genome Project [CONTINUED]

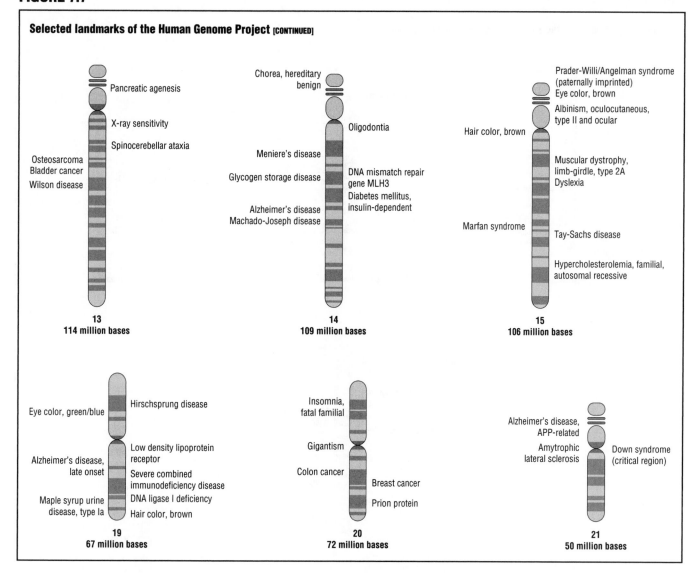

- Improved methods for the early detection of disease and to enable timely treatment when it is likely to be effective

- New technologies able to sequence the entire genome of any person affordably, ideally for less than $1,000

- Wider access to tools and technologies of "chemical genomics" to enhance understanding of biological pathways and accelerate pharmaceutical and other treatment research

Along with the special commemorative issue of *Nature*, the April 11, 2003, edition of *Science* ran articles that described the HGP and detailed the multidisciplinary DOE plan dubbed "Genomes to Life," which aimed to use HGP data to understand the ways in which microbes can provide opportunities to develop clean energy, reduce climate change, and clean the environment.

HGP REVISES ITS ESTIMATE

In October 2004 the NHGRI reduced its estimate of the number of human genes from 30,000 to 35,000 to between 20,000 and 25,000 (http://www.genome.gov/12513430). The refined human genome sequence, published in the October 21, 2004, issue of *Nature*, was the most complete version to date. It covered 99% of the gene-containing parts of the human genome, identified nearly all known genes, and was 99.9% accurate, according to the HGP scientists.

DAWN OF THE POSTGENOMIC ERA

The Human Genome Project Information Web site (October 9, 2006, http://www.ornl.gov/sci/techresources/Human_Genome/project/benefits.shtml) enumerates many of the potential benefits of HGP research. Besides its role in the practice of molecular medicine, other uses of HGP data and applications of human and other genomic research include:

- Microbial genomics—the use of bacteria to create new energy sources such as biofuels and safe, efficient toxic waste cleanup; enhanced understanding of how microbes cause disease; and protection from

FIGURE 7.7

Selected landmarks of the Human Genome Project [CONTINUED]

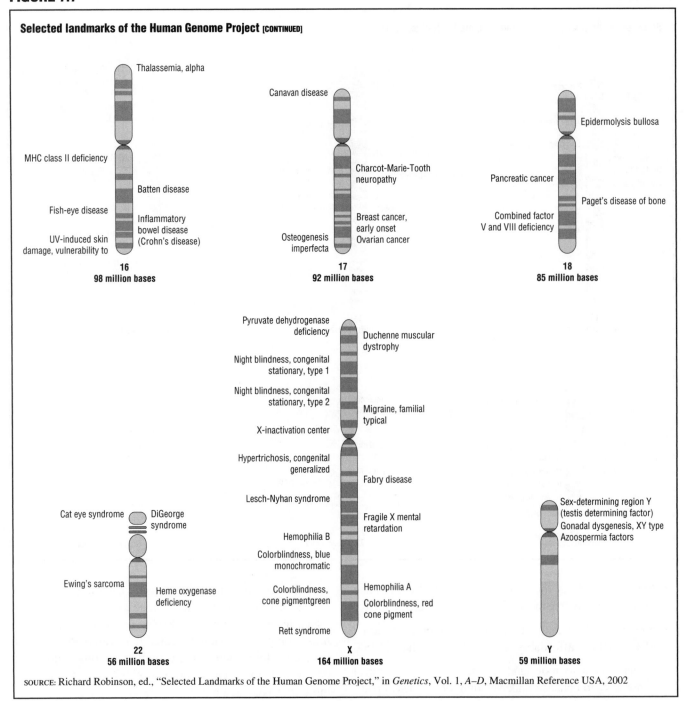

SOURCE: Richard Robinson, ed., "Selected Landmarks of the Human Genome Project," in *Genetics*, Vol. 1, *A–D*, Macmillan Reference USA, 2002

threats of biological and chemical terrorism and warfare (See Figure 7.8.)

- Risk assessment—measuring the risks and health problems caused by exposure to radiation, carcinogens (cancer-causing agents), and mutagenic chemicals; and reduction of the probability of heritable mutations

- Archaeology, anthropology, evolution, and human migration—comparing the genomes of humans and other organisms such as mice already has identified similar genes associated with diseases and traits; improving the understanding of germline (cells that

give rise to eggs or sperm) mutations; studying migration based on female genetic inheritance; examining mutations on the Y chromosome to trace lineage and migration of males; and comparing the DNA sequences of entire genomes of different microbes to enhance the understanding of the relationships among the three domains of life: archaebacteria (cells that do not contain nuclei), eukaryotes (cells that contain nuclei), and prokaryotes (single-celled organisms without nuclei)

- DNA forensics—identifying crime victims, potential suspects, and catastrophe victims through examination of DNA; confirm paternity and other family

FIGURE 7.8

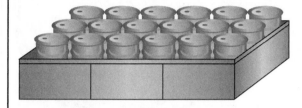

relationships; clear people wrongly accused of crimes; identify and protect endangered species; detect bacterial and other environmental pollutants; match organ donors and recipients for transplant programs; and determine pedigrees for animals and plants

- Agriculture and livestock breeding—develop healthier, stronger crops and farm animals able to resist insects, disease, and drought; create safer pesticides; grow more nutritious produce; incorporate vaccines into food products; and redeploy plants such as tobacco for use in environmental cleanup programs

Molecular Medicine

The HGP and the technological advances it has produced have moved the field of molecular medicine forward with extraordinary speed. In "Genomes, Transcriptomes, and Proteomes: Molecular Medicine and Its Impact on Medical Practice" (*Archives of Internal Medicine*, January 27, 2003), Ivan Gerling, Solomon S. Solomon, and Michael Bryer-Ash assert that the HGP will not only influence the way science is conducted but also will advance the clinical practice of medicine. Gerling, Solomon, and Bryer-Ash credit the HGP for the technological advances that enable preclinical detection—recognition of disease before its earliest biochemical or visible expression. They foresee increasing accuracy and ease of preclinical detection, as well as the ability to predict disease based on three fundamental levels of biologic determination:

- The genomic DNA constitution of the individual (the genome), which is unchanged from the moment of conception, except for some isolated, local mutations

- The transcribed messenger RNA complement (the transcriptome)

- The full range of translated proteins (the proteome)

Gerling, Solomon, and Bryer-Ash posit that the environment influences gene expression and modifies gene products in ways that initiate, accelerate, or slow progress of disease-causing processes. This does not change the genome, but it does change the transcriptome and the proteome. Recent technological breakthroughs have provided the tools to perform the comprehensive molecular analyses needed to examine not only the genome but also the transcriptome and proteome. Using new technologies will dramatically increase understanding at the molecular level of the mechanisms of disease development.

Along with molecular diagnosis of diseases even before they are clinically apparent, Gerling, Solomon, and Bryer-Ash predict increasingly effective therapies as genetic information enables physicians to individualize treatment in response to the availability of comprehensive genetic and molecular profiles. Although there are many promises and potential benefits of molecular medicine—improved diagnostic ease, speed, and accuracy; earlier detection of genetic predisposition or susceptibility to disease; gene therapy; and pharmaceutical drug development, specifically pharmacogenetics to produce "customized drugs"—Gerling, Solomon, and Bryer-Ash emphasize that the increased knowledge must be used responsibly. They caution that society must take steps to ensure that this improved understanding of genetics is not deployed to exclude people from obtaining insurance or employment.

Haplotype Mapping Project

In October 2002 an international effort to develop a haplotype map of the human genome was launched. A haplotype is a set of alleles or markers on one of a pair of homologous chromosomes, and a haplotype map will show human genetic variation. The premise of the International HapMap Project was that within the human genome different genetic variants within a chromosomal region—haplotypes—occur together far more frequently than others. Based on common haplotype patterns—combinations of DNA sequence variants that are usually found together—the haplotype map simplifies the search for medically important DNA sequence variations and offers new understanding of human population structure and history.

Given that any two people are 99.9% identical genetically, understanding the 0.1% difference is important because it helps explain why one person may be more susceptible to a certain disease than another. Researchers can use the HapMap to compare the genetic variation patterns of a group of people known to have a specific disease with a group of people without the disease. Finding a certain pattern more often in people with the disease identifies a genomic region that may contain genes that contribute to the condition. Researchers hope that identifying single nucleotide polymorphisms (SNPs), which are specific positions in the genome sequence that are occupied by one nucleotide in some copies and by a different nucleotide in others, will enable them to identify the alleles (particular forms of genes) that are associated with increased or decreased susceptibility to common diseases, such as asthma, heart disease, or psychiatric illness. Figure 7.9 shows that most SNP variation—about 85%—occurs within all populations.

Because investigators hypothesize that differences between haplotypes may be associated with varying susceptibility to disease, mapping the haplotype structure of the human genome may be the key to identifying the genetic basis of many common disorders. The HapMap project serves as a resource for studying the genetic factors that contribute to variation in response to environmental factors, in susceptibility to infection, and in the identification of genetic variants associated with the effectiveness of, and adverse responses to, drugs and vaccines.

To create a haplotype map, researchers must have enough SNPs to be sure that regions containing disease

FIGURE 7.9

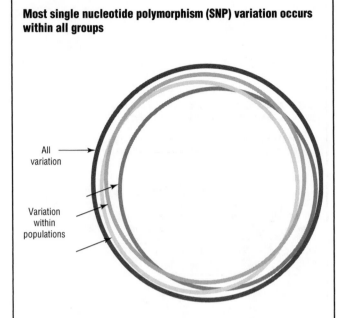

Most single nucleotide polymorphism (SNP) variation occurs within all groups

All variation

Variation within populations

SOURCE: "Most SNP Variation Is within All Groups," in *Developing a Haplotype Map of the Human Genome for Finding Genes Related to Health and Disease*, U.S. Department of Health and Human Services, National Institutes of Health, National Human Genome Research Institute, Division of Extramural Research, July 2001, http://www.genome.gov/10001665 (accessed October 27, 2006)

alleles have been found and that regions not containing disease alleles can be excluded from further consideration. The HapMap enables researchers to study the genetic risk factors underlying a wide range of disorders. For any given disease, researchers may perform an association study by using the HapMap tag SNPs to compare the haplotype patterns of a group of people known to have the disease to a group of people without the disease. If the association study finds a specific haplotype more frequently in those with the disease, researchers scrutinize the precise genomic region in their search for the specific genetic variant.

In the news release "Map of Human Genetic Variation Will Speed Search for Disease Genes" (February 7, 2005, http://www.nih.gov/news/pr/feb2005/nhgri-07.htm), the International HapMap Consortium announced plans to create an even more powerful map of human genetic variation than the group had initially intended. The project was originally intended to complete the map of haplotypes by September 2005, but by mid-2005 a draft of the HapMap, consisting of one million markers of genetic variation, was released. The first draft of the HapMap has enabled researchers to analyze the human genome in ways that were not possible with the human DNA sequence alone. Data from the second phase of the project were released in July 2006 and provide a denser map that enables scientists to narrow gene discovery more precisely to specific regions of the genome.

First Map of Common Human Genetic Variations

In February 2005 scientists working at Perlegen Sciences, Inc., in California produced the first map of common human genetic variations—differences in DNA that may assist in predicting disease risk and optimal disease treatment. To create the map, Perlegen investigators collaborated with researchers at the California Institute for Telecommunications and Information Technology at UC San Diego and the International Computer Science Institute at UC Berkeley. The map was unveiled at a meeting of the American Association for the Advancement of Science and described in the February 18, 2005, issue of *Science*.

Perlegen scientists looked at the DNA of seventy-one Americans of European, African, and Chinese ancestry and identified nearly 1.6 million SNPs—single-letter genetic differences—most of them shared across the three populations. Although the 1.6 million SNPs are just about 10% of the 10 million SNPs believed to exist, they appear to be among the most common. The Perlegen map does not pinpoint which SNPs are linked to disease risk, but future research will focus on identifying the SNP variations that trigger some people to develop diseases and others to resist or combat them.

Future of Genomic Research

In "A Vision for the Future of Genomics Research" (*Nature*, April 24, 2003), the NHGRI describes some of the research challenges of the postgenomic era. Besides the HapMap project and the DOE's "Genomes to Life," the article details other initiatives, including genome technology development.

Directed by the NHGRI, the Encyclopedia of DNA Elements (ENCODE) Project (December 2006, http://www.genome.gov/10005107) aims to develop efficient ways to identify and locate all the protein-coding genes, nonprotein-coding genes, and other sequence-based, functional elements contained in the human DNA sequence. This ambitious undertaking will produce an enormous resource for researchers seeking to use and apply the human sequence to predict disease risk and to develop new approaches to prevent and treat disease.

The ENCODE Project entails three phases: a three-year pilot project phase; a second technology development phase that parallels phase 1; and a planned production phase. In the October 22, 2004, issue of *Science*, the ENCODE investigators described their plans to build a "parts list" of all sequence-based functional elements in the human DNA sequence. The researchers hope to identify as-yet-unrecognized functional elements. During the pilot phase they are developing and testing high-throughput ways to efficiently identify functional elements. They are focusing on forty-four DNA targets, which together cover about 1% of the human genome, or about 30 million base pairs. The target regions were strategically selected to provide a representative cross section of the entire human genome sequence. In the second phase other researchers will work to develop new technologies to apply to the ENCODE Project. The results of the first two phases will determine how to begin the production phase and advance the ENCODE Project to analyze the remaining 99% of the human genome.

Another NHGRI initiative is the creation of publicly available libraries of organic chemical compounds for scientists engaged in charting biological pathways. These chemical compounds have many promising applications in genomic research. For example, their ability to enter cells readily makes them natural vehicles for pharmaceutical drug development and drug delivery system design. An endeavor of this size and scope requires significant financial and human resources, and the NHGRI is planning to use technologies such as robotic-enabled, high-output screening to create large libraries containing up to a million chemical compounds.

In 2004, the NIH announced the establishment of a Chemical Genomics Center based in the NHGRI Division of Intramural Research (http://www.ncgc.nih.gov/news/2004_06_09.html). Called PubChem (http://pubchem.ncbi.nlm.nih.gov/), it is a database of information on the biological activities of innovative chemical "tools" for use in biological research and drug development, and it includes a repository to acquire, maintain, and distribute a collection of up to one million chemical compounds. Like the HGP data, the chemical genomics network is freely available to the entire scientific community.

In addition, in April 2003 the United Kingdom's Wellcome Trust, along with Canadian funding organizations and the global pharmaceutical company GlaxoSmithKline, established a charitable organization, the Structural Genomics Consortium, to round out international efforts in structural genomics. Structural genomics is the systematic, high-volume generation of the three-dimensional structure of proteins. The goal of examining the structural genomics of any organism is the complete structural description of all proteins encoded by the genome of that organism. These descriptions are important for drug design, diagnosis, and treatment of disease. Like the HGP and Chemical Genomics Center, the Structural Genomics Consortium is placing all the protein structures in public databases where scientists throughout the world may access them.

CHAPTER 8

CLONING

The moral issues posed by human cloning are profound and have implications for today and for future generations. Today's overwhelming and bipartisan House action to prohibit human cloning is a strong ethical statement, which I commend. We must advance the promise and cause of science, but must do so in a way that honors and respects life.

—President George W. Bush, July 2001

We must not say to millions of sick or injured human beings "go ahead and die and stay paralyzed because we believe . . . a clump of cells is more important than you are."

—Representative Jerrold Nadler (D-NY), July 2001

The Human Genome Project defines three distinct types of cloning. The first is the use of highly specialized deoxyribonucleic acid (DNA) technology to produce multiple, exact copies of a single gene or other segment of DNA to obtain sufficient material to examine for research purposes. This process produces cloned collections of DNA known as clone libraries. The second kind of cloning involves the natural process of cell division to create identical copies of the entire cell. These copies are called a cell line. The third type of cloning, reproductive cloning, is the one that has received the most attention in the mass media. This is the process that generates complete, genetically identical organisms such as Dolly, the famous Scottish sheep cloned in 1996 and named after the entertainer Dolly Parton.

Cloning may also be described by the technology used to perform it. For example, the term *recombinant DNA technology* describes the technology and mechanism of DNA cloning. Also known as molecular cloning, or gene cloning, it involves the transfer of a specific DNA fragment of interest to researchers from one organism to a self-replicating genetic element of another species such as a bacterial plasmid. (See Figure 8.1.) The DNA under study may then be reproduced in a host cell. This technology has been in use since the 1970s and is a standard practice in molecular biology laboratories.

Just as GenBank is an online public repository of the human genome sequence, the Clone Registry database is a sort of "public library." Used by genome sequencing centers to record which clones have been selected for sequencing, which sequencing efforts are currently under way, and which are finished and represented by sequence entries in GenBank, the Clone Registry may be freely accessed by scientists worldwide. To effectively coordinate all of this information, a standardized system of naming clones is essential. The nomenclature used is shown in Figure 8.2.

CLONING GENES

Molecular cloning is performed to enable researchers to have many copies of genetic material available in the laboratory for the purpose of experimentation. Cloned genes allow researchers to examine encoded proteins and are used to sequence DNA. Gene cloning also allows researchers to isolate and experiment on the genes of an organism. This is particularly important in terms of human research; in instances where direct experimentation on humans might be dangerous or unethical, experimentation on cloned genes is often practical and feasible.

Cloned genes are also used to produce pharmaceutical drugs, insulin, clotting factors, human growth hormone, and industrial enzymes. Before the widespread use of molecular cloning, these proteins were difficult and expensive to manufacture. For example, before recombinant DNA technology, insulin (a pancreatic hormone that regulates blood glucose levels) used by people with diabetes was extracted and purified from cow and pig pancreases. Because the amino acid sequences of insulin from cows and pigs are slightly different from those in human insulin, some patients experienced adverse immune reactions to the nonhuman "foreign insulin."

FIGURE 8.1

Bacterial plasmid

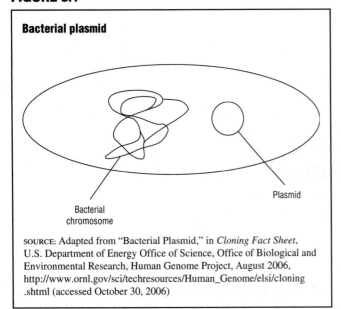

Bacterial chromosome

Plasmid

SOURCE: Adapted from "Bacterial Plasmid," in *Cloning Fact Sheet*, U.S. Department of Energy Office of Science, Office of Biological and Environmental Research, Human Genome Project, August 2006, http://www.ornl.gov/sci/techresources/Human_Genome/elsi/cloning.shtml (accessed October 30, 2006)

FIGURE 8.2

Standardized clone names

A clone is identified by its microtitre plate address (plate number, row, and column) and prefixed by a library abbreviation to produce a unique name.

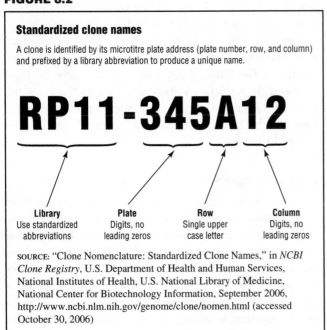

RP11-345A12

Library	Plate	Row	Column
Use standardized abbreviations	Digits, no leading zeros	Single upper case letter	Digits, no leading zeros

SOURCE: "Clone Nomenclature: Standardized Clone Names," in *NCBI Clone Registry*, U.S. Department of Health and Human Services, National Institutes of Health, U.S. National Library of Medicine, National Center for Biotechnology Information, September 2006, http://www.ncbi.nlm.nih.gov/genome/clone/nomen.html (accessed October 30, 2006)

The recombinant human version of insulin is identical to human insulin so it does not produce an immune reaction.

Figure 8.3 shows how a gene is cloned. First, a DNA fragment containing the gene being studied is isolated from chromosomal DNA using restriction enzymes. It is joined with a plasmid (a small ring of DNA found in many bacteria that can carry foreign DNA) that has been cut with the same restriction enzymes. When the fragment of chromosomal DNA is joined with its cloning vector (cloning vectors, such as plasmids and yeast artificial chromosomes, introduce foreign DNA into host cells), it is called a recombinant DNA molecule. Once it

FIGURE 8.3

Cloning DNA in plasmids

By fragmenting DNA of any origin (human, animal, or plant) and inserting it in the DNA of rapidly reproducing foreign cells, billions of copies of a single gene or DNA segment can be produced in a very short time. DNA to be cloned is inserted into a plasmid (a small, self-replicating circular molecule of DNA) that is separate from chromosomal DNA. When the recombinant plasmid is introduced into bacteria, the newly inserted segment will be replicated along with the rest of the plasmid.

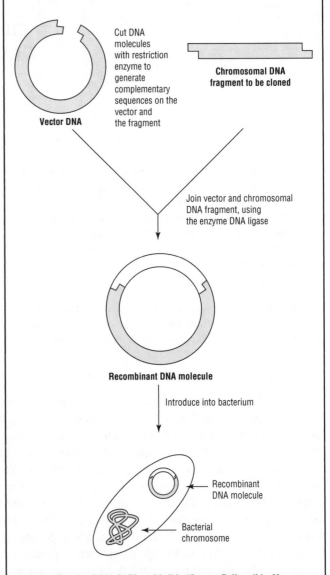

Cut DNA molecules with restriction enzyme to generate complementary sequences on the vector and the fragment

Vector DNA

Chromosomal DNA fragment to be cloned

Join vector and chromosomal DNA fragment, using the enzyme DNA ligase

Recombinant DNA molecule

Introduce into bacterium

Recombinant DNA molecule

Bacterial chromosome

SOURCE: "Cloning DNA in Plasmids," in "Image Gallery," in *Human Genome Project Information*, U.S. Department of Energy Office of Science, Office of Biological and Environmental Research, Human Genome Project, http://www.ornl.gov/sci/techresources/Human_Genome/graphics/slides/ttplasmid.shtml (accessed December 15, 2006)

has entered into the host cells, the recombinant DNA can be reproduced along with the host cell DNA.

Another molecular cloning technique that is simpler and less expensive than the recombinant cloning method is the polymerase chain reaction (PCR). PCR has also been dubbed "molecular photocopying" because it amplifies DNA without the use of a plasmid. Figure 6.5 in chapter 6 shows how PCR is used to generate a virtually unlimited number of copies of a piece of DNA.

REPRODUCTIVE CLONING

Another way to describe and classify cloning is by its purpose. Organismal or reproductive cloning is a technology used to produce a genetically identical organism—an animal with the same nuclear DNA as an existing, or even an extinct, animal.

The reproductive cloning technology used to create animals is called somatic cell nuclear transfer (SCNT). In SCNT scientists transfer genetic material from the nucleus of a donor adult cell to an enucleated egg (an egg from which the nucleus has been removed). This eliminates the need for fertilization of an egg by a sperm. The reconstructed egg containing the DNA from a donor cell is treated with chemicals or electric current to stimulate cell division. Once the cloned embryo reaches a suitable stage, it is transferred to the uterus of a surrogate (female host), where it continues to grow and develop until birth. Figure 8.4 shows the entire SCNT process that culminates in the transfer of the embryo into the surrogate mother.

Organisms or animals generated using SCNT are not perfect or identical clones of the donor organism or "parent" animal. Although the clone's nuclear DNA is identical to the donor's, some of the clone's genetic materials come from the mitochondria in the cytoplasm of the enucleated egg. Mitochondria, the organelles that serve as energy sources for the cell, contain their own short segments of DNA called mtDNA. Acquired mutations in the mDNA contribute to differences between clones and their donors and are believed to influence the aging process.

Dolly the Sheep Paves the Way for Other Cloned Animals

In 1952 scientists transferred a cell from a frog embryo into an unfertilized egg, which then developed into a tadpole. This process became the prototype for cloning. Ever since, scientists have been cloning animals. The first mammals were also cloned from embryonic cells in the 1980s. In 1997 cloning became headline news when, following more than 250 failed attempts, Ian Wilmut and his colleagues at the Roslin Institute in Edinburgh, Scotland, successfully cloned a sheep, which they named Dolly. Dolly was the first mammal cloned from the cell of an adult animal, and since then researchers have used cells from adult animals and various modifications of nuclear transfer technology to clone a range of animals, including sheep, goats, cows, mice, pigs, cats, and rabbits.

To create Dolly, the Roslin Institute researchers transplanted a nucleus from a mammary gland cell of a Finn Dorsett sheep into the enucleated egg of a Scottish blackface ewe and used electricity to stimulate cell division. The newly formed cell divided and was placed in the uterus of a

FIGURE 8.4

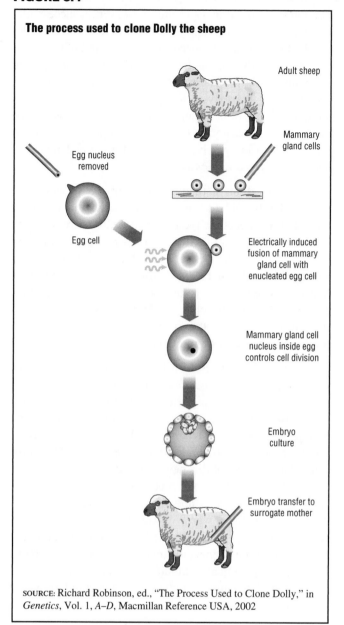

The process used to clone Dolly the sheep

Adult sheep

Mammary gland cells

Egg nucleus removed

Egg cell

Electrically induced fusion of mammary gland cell with enucleated egg cell

Mammary gland cell nucleus inside egg controls cell division

Embryo culture

Embryo transfer to surrogate mother

SOURCE: Richard Robinson, ed., "The Process Used to Clone Dolly," in *Genetics*, Vol. 1, *A–D*, Macmillan Reference USA, 2002

blackface ewe to gestate. Born several months later, Dolly was a true clone—genetically identical to the Finn Dorsett mammary cells and not to the blackface ewe, which served as her surrogate mother. Her birth revolutionized the world's understanding of molecular biology, ignited worldwide discussion about the morality of generating new life through cloning, prompted legislation in dozens of countries, and launched an ongoing political debate in the U.S. Congress.

Dolly was the object of intense media and public fascination. She proved to be a basically healthy clone and produced six lambs of her own through normal sexual means. Before her death by lethal injection on February 14, 2003, Dolly had been suffering from lung cancer and arthritis. An autopsy (postmortem examination) of Dolly revealed that, other than her cancer and arthritis, she was anatomically like other sheep.

In February 1997 Don Wolf and his colleagues at the Oregon Regional Primate Center in Beaverton successfully cloned two rhesus monkeys using laboratory techniques that had previously produced frogs, cows, and mice. It was the first time that researchers used a nuclear transplant to generate monkeys. The monkeys were created using different donor blastocysts (early-stage embryos), so they were not clones of one another—each monkey was a clone of the original blastocyst that had developed from a fertilized egg. Unfortunately, neither of the cloned monkeys survived past the embryonic stage.

An important distinction between the process that created Dolly and the one that produced the monkeys was that unspecialized embryonic cells were used to create the monkeys, whereas a specialized adult cell was used to create Dolly. The Oregon experiment was followed closely in the scientific and lay communities because, in terms of evolutionary biology and genetics, primates are closely related to humans. Researchers and the public speculated that if monkeys could be cloned, it might become feasible to clone humans.

In May 2001 BresaGen Limited, an Australian biotechnology firm, announced the birth of that country's first cloned pig. The pig was cloned from cells that had been frozen in liquid nitrogen for more than two years, and the company used technology that was different from the process used to clone Dolly the sheep. The most immediate benefit of this new technology was to improve livestock—cloning enables breeders to take some animals with superior genetics and rapidly produce more. Biomedical scientists were especially attentive to this research because of its potential for xenotransplantation—the use of animal organs for transplantation into humans. Pig organs genetically modified so that they are not rejected by the human immune system could prove to be a boon to medical transplantation.

During the same year the first cat was cloned, and the following year rabbits were successfully cloned. In January 2003 researchers at Texas A&M University reported that cloned pigs behaved normally—as expected for a litter of pigs—but were not identical to the animals from which they were cloned in terms of food preferences, temperament, and how they spent their time. The researchers explained the variation as arising from the environment and epigenetic (not involving DNA sequence change) factors, causing the DNA to line up differently in the clones. Epigenetic activity is defined as any gene-regulating action that does not involve changes to the DNA code and that persists through one or more generations, and it may explain why abnormalities such as fetal death occur more frequently in cloned species.

On May 4, 2003, a cloned mule—the first successful clone of any member of the horse family—was born in Idaho. The clone was not just any mule, but the brother of the world's second fastest racing mule. Named Idaho Gem, the cloned mule was created by researchers at the University of Idaho and Utah State University. The researchers attributed their success to changes in the culture medium they used to nurture the eggs and embryos.

In August 2003 scientists at the Laboratory of Reproductive Technology in Cremona, Italy, were the first to clone a horse. The Italian scientists, Cesare Galli et al., describe their cloning technique in "Pregnancy: A Cloned Horse Born to Its Dam Twin" (*Nature*, August 7, 2003).

The mule was cloned from cells extracted from a mule fetus, whereas the cloned horse's DNA came from her adult mother's skin cells. There were other differences as well. The University of Idaho and Utah State University researchers harvested fertile eggs from mares, removed the nucleus of each egg, and inserted DNA from cells of a mule fetus. The reconstructed eggs were then surgically implanted into the wombs of female horses. In contrast, the Italian scientists harvested hundreds of eggs from mare carcasses, cultured the eggs, removed their DNA, and replaced it with DNA taken from either adult male or female horse skin cells.

In May 2004 the first bull was cloned from a previously cloned bull in a process known as serial somatic cell cloning or recloning. Before the bull, the only other successful recloning efforts involved mice. Chikara Kubota, X. Cindy Tian, and Xiangzhong Yang, the successful research team, describe their techniques in "Serial Bull Cloning by Somatic Cell Nuclear Transfer" (*Nature Biotechnology*, May 23, 2004). Their effort was also cited in the *Guinness Book of World Records* as the "largest clone in the world."

At the close of 2004 a South Korean research team reported cloning macaque monkey embryos, which would be used as a source of stem cells. Conservationists then focused research efforts on cloning rare and endangered species. In April 2005 Texas A&M University announced the first successfully cloned foal in the United States. That same month, Korean scientists at Seoul National University (SNU) cloned a dog they dubbed "Snuppy." In May 2005 the Brazilian Agricultural Research Corporation, Embrapa, reported the creation of two cloned calves from a Junquiera cow, which is an endangered species.

Cloning Endangered Species

Reproductive cloning technology may be used to repopulate endangered species such as the African bongo antelope, the Sumatran tiger, and the giant panda, or animals that reproduce poorly in zoos or are difficult to breed. On January 8, 2001, scientists at Advanced Cell Technology (ACT), a biotechnology company in Massachusetts, announced the birth of the first clone of an endangered animal, a baby bull gaur—a large wild ox from India and Southeast Asia—named Noah. Noah was cloned using the nuclei of frozen skin cells taken from an adult male gaur that had died eight years earlier. The skin cell nuclei were joined with enucleated cow eggs, one of which was implanted into a surrogate cow mother. Unfortunately, the cloned gaur died from an infection within days of its birth. The same year scientists in Italy successfully cloned an endangered wild sheep. Cloning an endangered animal is different from cloning a more common animal because cloned animals need surrogate mothers to be carried to term. The transfer of embryos is risky, and researchers are reluctant to put an endangered animal through the rigors of surrogate motherhood, opting to use nonendangered domesticated animals whenever possible.

Cloning extinct animals is even more challenging than cloning living animals because the egg and the surrogate mother used to create and harbor the cloned embryo are not the same species as the clone. Furthermore, for most already extinct animal species such as the woolly mammoth or dinosaur, there is insufficient intact cellular and genetic material from which to generate clones. In the future, carefully preserving intact cellular material of imperiled species may allow for their preservation and propagation.

In April 2003 ACT announced the birth of a healthy clone of a Javan banteng, an endangered cattlelike animal native to Asian jungles. The clone was created from a single skin cell, taken from another banteng before it died in 1980, which had remained frozen until it was used to create the clone. The banteng embryo gestated in a standard beef cow in Iowa.

Born April 1, 2003, the cloned banteng developed normally, growing its characteristic horns and reaching an adult weight of about 1,800 pounds. He was nicknamed Stockings and, as of 2007, lived at the San Diego Zoo. Hunting and habitat destruction have reduced the number of banteng, which once lived in large numbers in the bamboo forests of Asia, by more than 75% from 1983 to 2003.

In August 2005 the Audubon Nature Institute in New Orleans, Louisiana, reported that two unrelated endangered African wildcat clones had given birth to eight babies. Their births confirmed that clones of wild animals can breed naturally, which is vitally important for protecting endangered animals on the brink of extinction.

Reproductive Human Cloning

In December 2002 a religious sect known as the Raelians made news when their private biotechnology firm, Clonaid, announced that they had successfully delivered "the world's first cloned baby." The announcement, which could not be independently verified or substantiated, generated unprecedented media coverage and was condemned in the scientific and lay communities. At least some of the media frenzy resulted from the beliefs of the Raelians—namely, the sect contends that humans were created by extraterrestrial beings. In 2005 Clonaid claimed to have produced at least thirteen cloned children, but as of 2007 had not yet offered any proof of their existence.

Clonaid's announcement brought attention to the fact that several laboratories around the world had embarked on clandestine efforts to clone a human embryo. For example, in 2002 a U.S. fertility specialist, Panayiotis Zavos, claimed to be collaborating with about two dozen international researchers to produce human clones. Another doctor focusing on fertility issues, Severino Antinori, attracted media attention when he maintained that hundreds of infertile couples in Italy and thousands in the United States had already enrolled in his human cloning initiative. Neither these researchers nor anyone else had offered proof of successful reproductive human cloning as of early 2007.

THERAPEUTIC CLONING

Therapeutic cloning (also called embryo cloning) is the creation of embryos for use in biomedical research. The objective of therapeutic cloning is not to create clones but to obtain stem cells. Stem cells are "master cells" capable of differentiating into multiple other cell types. This potential is important to biomedical researchers because stem cells may be used to generate any type of specialized cell, such as nerve, muscle, blood, or brain cells. Many scientists believe that stem cells can not only provide a ready supply of replacement tissue but also may hold the key to developing more effective treatments for common disorders such as heart disease and cancer as well as degenerative diseases such as Alzheimer's and Parkinson's. Researchers believe that in the future it may be possible to induce stem cells to grow into complete organs.

Advocates of therapeutic cloning point to other treatment benefits such as using stem cells to generate bone marrow for transplants. They contend that scientists could use therapeutic cloning to manufacture perfectly matched bone marrow using the patient's own skin or other cells. This would eliminate the problem of rejection of foreign tissue associated with bone marrow transplant and other organ transplantation. Stem cells also have the potential to repair and restore damaged heart and nerve tissue. Furthermore, there is mounting evidence to suggest that stem cells from cloned embryos have greater potential as medical treatments than stem cells harvested from unused embryos at fertility clinics, which are created by in vitro fertilization and are now the major source of stem cells for research. These prospective benefits are among the most compelling arguments in favor of cloning to obtain embryonic stem cells.

Stem cells used in research are harvested from the blastocyst after it has divided for five days, during the earliest stage of embryonic development. Many people regard human embryos as human beings or at least potential human beings and consider their destruction, or even using techniques to obtain stem cells that might imperil their future viability, as immoral or unethical.

In November 2001 the ACT researchers Jose B. Cibelli et al. reported in "Somatic Cell Nuclear Transfer in Humans: Pronuclear and Early Embryonic Development" (*e-biomed: The Journal of Regenerative Medicine*, November 26, 2001) that they had created a cloned human embryo, and, unlike groups that had claimed to have done this before, they published their results. The ACT press release "Advanced Cell Technology, Inc. (ACT) Today Announced Publication of Its Research on Human Somatic Cell Nuclear Transfer and Parthenogenesis" (November 25, 2001, http://www.advancedcell .com) boasted that this achievement offered "the first proof that reprogrammed human cells can supply tissue" and asserted that this accomplishment was a vital first step toward the objective of therapeutic cloning—using cloned embryos to harvest embryonic stem cells able to grow into replacement tissue perfectly matched to individual patients. To clone the human embryos, Cibelli et al. collected women's eggs and painstakingly removed the genetic material from the eggs with a thin needle. A skin cell was inserted inside each of eight enucleated eggs, which were then chemically stimulated to divide. Just three of the eight eggs began dividing, and only one reached six cells before cell division ceased.

That same year investigators at the South Australian Research and Development Institute used lambs to experiment with therapeutic cloning. The goal was to replace cells stricken with Parkinson's disease with healthy ones derived from a cloned embryo. In 2003 researchers in Italy reported successfully using adult stem cells to cure mice that had a form of multiple sclerosis. The scientists injected the diseased mice with stem cells that had been extracted from the brains of adult mice reproduced in the laboratory. Postmortem examination of the mice showed that the stem cells had migrated to and then repaired damaged areas of the nerves and brain.

In August 2003 a Chinese research team led by Huizhen Sheng, an American-trained scientist working at the Shanghai Second Medical University, reported that it had made human embryonic stem cells by combining human skin cells with rabbit eggs. Their accomplishment was published in the Chinese scientific journal *Cell Research*, a peer-reviewed publication of the Shanghai Institute of Cell Biology and the Chinese Academy of Sciences. The researchers removed the rabbit eggs' DNA and injected human skin cells inside them. The eggs then grew to form embryos containing human genetic material. After several days the embryos were dissected to extract their stem cells.

In February 2004 scientists at Seoul National University in South Korea reported in the journal *Science* that they had successfully cloned healthy human embryos, removed embryonic stem cells, and grown them in mice. In January 2006, following a lengthy investigation, Seoul National University concluded that the research reported in *Science* had been fabricated. As a result, the journal retracted the article along with another study by the same author. In May 2006, the investigator, Hwang Woo-suk, was charged with fraud, embezzlement, and violating South Korea's bioethics statutes.

In 2005 Wilmut was granted a license by the British government to clone human embryos to generate stem cell lines to study motor neuron disease (MND). Wilmut and his colleagues are working to clone embryos to generate stem cells that would in turn become motor neurons with MND-causing gene defects. By observing the stem cells grow into neurons, the researchers hope to discover what causes the cells to degenerate. Their research involves comparing the stem cells with healthy and diseased cells from MND patients to gain a better understanding of the illness and to test potential drug treatments.

Human reproductive cloning remains illegal in Britain but therapeutic cloning—creating embryos as a source of stem cells to cure diseases—is allowed on an approved basis. The license granted to Wilmut and his colleagues is the second one granted by Britain's Human Fertilisation and Embryology Authority.

In July 2006 the researchers Deepa Deshpande et al. restored movement to paralyzed rats using a new method that demonstrates the potential of embryonic stem cells to restore function to humans suffering from neurological disorders. They published their results in "Recovery from Paralysis in Adult Rats Using Embryonic Stem Cells" (*Annals of Neurology*, July 2006). Although clinical trials in humans are still years away, the results of this research represent an important advance in the quest for a cure for paralysis and other neurological disorders.

In October 2006 Kevin A. D'Amour et al., in "Production of Pancreatic Hormone-Expressing Endocrine Cells from Human Embryonic Stem Cells" (*Nature Biotechnology*, October 19, 2006), reported developing a process to turn human embryonic stem cells into pancreatic cells that can produce insulin and other hormones. The researchers anticipate testing these cells in animals in 2008 and if the animal studies are successful, then clinical trials in human patients may begin as soon as 2009.

Three studies—Volker Schächinger et al. in "Intracoronary Bone Marrow-Derived Progenitor Cells in Acute Myocardial Infarction," Ketil Lunde et al. in "Intracoronary Injection of Mononuclear Bone Marrow Cells in Acute Myocardial Infarction," and Birgit Assmus et al. in "Transcoronary Transplantation of Progenitor Cells after Myocardial Infarction"—describing the use of stem cells in the treatment of heart disease were published in the September 21, 2006, issue of the *New England Journal of Medicine*. The studies produced conflicting results: Schächinger and his colleagues reported benefits for patients who had suffered myocardial infarction (heart attack). Lunde and his contributors found no benefit from stem cell treatment of such patients. Assmus and her collaborators studied patients with chronic heart failure, who did show improvement after treatment. In the editorial "Cardiac Cell Therapy—Mixed Results from Mixed Cells" in the same issue of the journal, Antony Rosenzweig writes that the three studies "provide a realistic perspective on this approach while leaving room for cautious optimism and underscoring the need for further study."

Rick Weiss, in "Stem Cell Work Shows Promise and Risks" (*Washington Post*, October 23, 2006), reports that research conducted at the University of Rochester Medical Center using nerve cells grown from human embryonic stem cells to treat rats afflicted with Parkinson's disease produced mixed results. The treatment reduced the animals' symptoms, but caused tumors in the rodents' brains. The researchers acknowledged that their work showed both the promise and risks associated with stem cell treatments.

Research Promises Therapeutic Benefits without Cloning

In "Homologous Recombination in Human Embryonic Stem Cells" (*Nature Biotechnology*, March 2003), Thomas P. Zwaka and James A. Thomson report that they used human embryonic stem cells to splice out individual genes and substituted different genes in their place. Their accomplishment was heralded as a first step toward the goal of regenerating parts of the human body by transplanting either stem cells or tissues grown from stem cells into patients. Zwaka and Thomson used electrical charges and chemicals to make the cells' membranes permeable; the cells allowed the customized genes to enter, and they then found and replaced their counterparts in the cells' DNA.

The ability to make precise genetic changes in human stem cells could be used to boost their therapeutic potential or make them more compatible with patients' immune systems. Some researchers assert that the success of this bioengineering feat might eliminate the need to pursue the hotly debated practice of therapeutic cloning,

but others caution that such research could heighten concerns among those who fear that stem cell technology will lead to the creation of "designer babies," which are bred for specific characteristics such as appearance, intelligence, or athletic prowess.

In May 2003 the University of Pennsylvania researcher Hans R. Schöler and his colleagues announced another historic first: The researchers transformed ordinary mouse embryo cells into egg cells in laboratory dishes ("Scientists Produce Mouse Eggs from Embryonic Stem Cells, Demonstrating Totipotency Even In Vitro," *ScienceDaily*, May 2, 2003). Schöler selected from a population of stem cells the ones that bore certain genetic traits suggesting the potential to become eggs. They then isolated those in laboratory dishes. Eventually, the cells morphed into two kinds of cells, including young egg cells. The eggs matured normally and appeared to be healthy in terms of their appearance, size, and gene expression. When cultured for a few days, the eggs also underwent spontaneous division and formed structures resembling embryos, a process called parthenogenesis. This finding implies that the eggs were fully functional and likely could be fertilized with sperm.

Once refined, this technology could be applied to produce egg cells in the laboratory that would enable scientists to engineer traits into animals and help conservationists rebuild populations of endangered species. It offers researchers the chance to observe mammalian egg cells as they mature, a process that occurs unseen within the ovary. The technology also offers an unparalleled opportunity to learn about meiosis (reduction division), the process of cell division during which an egg or sperm disgorges half of its genes so it can join with a gamete of the opposite sex. There are many potential medical benefits as well. For example, women who cannot make healthy eggs could use this technology to ensure healthy offspring.

Like many new technologies, transforming cells into eggs simultaneously resolves existing ethical issues and creates new ones. For example, because the embryonic stem cells spontaneously transformed themselves into eggs, this procedure overcomes many of the ethical objections to cloning, which involves creating offspring from a single parent. However, it also paves the way for the creation of "designer eggs" from scratch and, if performed with human cells, could redefine the biological definitions of mothers and fathers.

In September 2003 efforts to transform stem cells into sperm were successful. Toshiaki Noce and his colleagues in Tokyo, Japan, observed male mouse embryonic stem cells that developed spontaneously, with some cells actually becoming germ cells. When the researchers transplanted the germ cells into testicular tissue, the cells underwent meiosis and formed sperm cells. One possible medical application of this technology would be to assist couples who are infertile because the male cannot produce healthy sperm. One of the ethical issues that might arise would be the potential for two men to both be biological fathers of a child. Another is the potential to generate a human being who never had any parents using two laboratory-grown stem cells, one transformed into a sperm and the other into an egg. Many ethicists advise consideration of such issues before permitting human experimentation.

In 2004 the National Institutes of Health (NIH) reported that researchers from the University of Pennsylvania School of Veterinary Medicine used cells from mice to grow sperm progenitor cells in a laboratory culture (November 3, 2004, http://www.nih.gov/news/pr/nov2004/nichd-03.htm). Known as spermatogonial stem cells, the progenitor cells are incapable of fertilizing egg cells but give rise to cells that develop into sperm. The researchers transplanted the cells into infertile mice, which were then able to produce sperm and father offspring that were genetically related to the donor mice.

This breakthrough has many potential applications, including developing new treatments for male infertility and extending the reproductive lives of endangered species. Researchers will also attempt to genetically manipulate the sperm cells grown in a culture medium and then implant the cells into animals. In this way they could introduce new traits into laboratory animals and livestock, such as disease resistance. The culture technique offers researchers additional opportunities to investigate the potential of spermatogonial stem cells as a source for adult stem cells to replace diseased or injured tissue.

New Methods of Obtaining Stem Cells without Destroying Embryos

In "Embryonic and Extraembryonic Stem Cell Lines Derived from Single Mouse Blastomeres" (*Nature*, January 12, 2006), Young Chung et al. report that embryonic stem cell cultures could be derived from single cells of mouse embryos. Irina Klimanskaya et al., in "Human Embryonic Stem Cell Lines Derived from Single Blastomeres" (*Nature*, August 23, 2006), describe a technique for removing a single cell—called a blastomere—from a three-day-old embryo with eight to ten cells and using a biochemical process to create embryonic stem cells from the blastomere. The method of removing a cell from the embryo is much like the technique used for preimplantation genetic diagnosis, which is performed to screen the cell for genetic defects. The researchers note that human embryonic stem cell lines derived from a single blastomere were comparable to lines derived with conventional techniques. Although Klimanskaya and her colleagues assert that the new method "will make it far more difficult to oppose this research," opponents of stem cell research contend that the new technique is morally unacceptable because even a single cell removed from an early embryo may have the potential to produce a life.

Another technique reported in 2006 can obviate the need for embryonic stem cells. Erika Check notes in "Simple Recipe Gives Adult Cells Embryonic Powers" (*Nature*, July 6, 2006) that researchers in the United Kingdom discovered the gene, called nanog, that is the key to "reprogramming" adult cells back to an embryonic state. The reprogramming of adult cells using nanog may make it possible for scientists to generate cells that specialize and develop into every type of cell in the body without the controversial use of human embryonic stem cells.

OPINIONS SHAPE PUBLIC POLICY

The difficulty and low success rate of much animal reproductive cloning (an average of just one or two viable offspring result from every one hundred attempts) and the as-yet-inadequate understanding about reproductive cloning have prompted many scientists to deem it unethical to attempt to clone humans. Many attempts to clone mammals have failed, and about one-third of clones born alive suffer from anatomical, physiological, or developmental abnormalities that are often debilitating. Some cloned animals have died prematurely from infections and other complications at rates higher than conventionally bred animals, and some researchers anticipate comparable outcomes from human cloning. Furthermore, scientists cannot yet describe or characterize how cloning influences intellectual and emotional development. Even though the attributes of intelligence, temperament, and personality may not be as important for cattle or other primates, they are vital for humans. Without considering the myriad religious, social, and other ethical concerns, the presence of so many unanswered questions about the science of reproductive cloning has prompted many investigators to consider any attempts to clone humans as scientifically irresponsible, unacceptably risky, and morally unallowable.

On August 9, 2001, President George W. Bush (http://www.whitehouse.gov/news/releases/2001/08/20010809-2.html) announced his decision to allow federal funds to be used for research on existing human embryonic stem cell lines as long as the derivation process (which begins with the removal of the inner cell mass from the blastocyst) had already been initiated and the embryo from which the stem cell line was derived no longer had the possibility of development as a human being. The president established the following criteria that research studies must meet to qualify for federal funding:

- The stem cells must have been drawn from an embryo created for reproductive purposes that was no longer needed for these purposes.

- Informed consent must have been obtained for the donation of the embryo and no financial inducements provided for donation of the embryo.

In January 2002 the Panel on Scientific and Medical Aspects of Human Cloning was convened by the National Academy of Sciences; the National Academy of Engineering; the Institute of Medicine Committee on Science, Engineering, and Public Policy; and the National Research Council, Division on Earth and Life Studies Board on Life Sciences. Following the panel, the report *Scientific and Medical Aspects of Human Cloning* (January 2002, http://www7.nationalacademies.org/cosepup/Human_Cloning.html) was issued that called for a ban on human reproductive cloning.

The panel recommended a legally enforceable ban with substantial penalties as the best way to discourage human reproductive cloning experiments in both the public and private sectors. It cautioned that a voluntary measure might be ineffective because many of the technologies needed to accomplish human reproductive cloning are widely accessible in private clinics and other organizations that are not subject to federal regulations.

The panel did not, however, conclude that the scientific and medical considerations that justify a ban on human reproductive cloning are applicable to nuclear transplantation to produce stem cells. In view of their potential to generate new treatments for life-threatening diseases and advance biomedical knowledge, the panel recommended that biomedical research using nuclear transplantation to produce stem cells be permitted. Finally, the panel encouraged ongoing national discussion and debate about the range of ethical, societal, and religious issues associated with human cloning research.

On February 14, 2002, the American Association for the Advancement of Science (AAAS; http://archives.aaas.org/docs/documents.php?doc_id=425), the world's largest general scientific organization, affirmed a legally enforceable ban on reproductive cloning; however, the AAAS supported therapeutic or research cloning using nuclear transplantation methods under appropriate government oversight. Similarly, the American Medical Association (AMA; April 6, 2006, http://www.ama-assn.org/ama/pub/category/4560.html), a national physicians' organization, issued a formal public statement against human reproductive cloning. The AMA statement cautioned that human cloning failures could jeopardize promising science and genetic research and prevent biomedical researchers and patients from realizing the potential benefits of therapeutic cell cloning.

On April 10, 2002, President Bush called on the Senate to back legislation banning all types of human cloning (http://www.whitehouse.gov/news/releases/2002/04/20020410-4.html). In his plea to the Senate, Bush said:

Science has set before us decisions of immense consequence. We can pursue medical research with a clear sense of moral purpose or we can travel without an ethical compass into a world we could live to regret. Science now presses forward the issue of human cloning. How we answer the question of human cloning will place us on one path or the other. . . . Human cloning is deeply troubling to me, and to most Americans. Life is a creation, not a commodity. Our children are gifts to be loved and protected, not products to be designed and manufactured. Allowing cloning would be taking a significant step toward a society in which human beings are grown for spare body parts, and children are engineered to custom specifications; and that's not acceptable. . . . I believe all human cloning is wrong, and both forms of cloning ought to be banned, for the following reasons. First, anything other than a total ban on human cloning would be unethical. Research cloning would contradict the most fundamental principle of medical ethics, that no human life should be exploited or extinguished for the benefit of another.

On September 25, 2002, Elias Zerhouni, the director of the NIH, testified before the Senate Appropriations Subcommittee on Labor, Health and Human Services, and Education in favor of advancing the field of stem cell research (http://olpa.od.nih.gov/hearings/107/session2/testimonies/stemcelltest.asp). Zerhouni exhorted Congress to continue to fund both human embryonic stem cell research and adult stem cell research simultaneously to learn as much as possible about the potential of both types of cells to treat human disease. He observed that many studies that do not involve human subjects must be performed before any new therapy is tested on human patients. These preclinical studies include tests of the long-term survival and fate of transplanted cells, as well as tests of the safety, toxicity, and effectiveness of the cells in treating specific diseases in animals. Zerhouni promised that trials using human subjects, the clinical research phase, would begin only after the basic foundation had been established. Despite Zerhouni's impassioned plea and subsequent efforts to advance stem cell research, at the close of 2006 U.S. law continued to ban federal funding of any research that might harm human embryos.

Moral and Ethical Objections to Human Cloning

People who oppose human cloning are as varied as the interests and institutions they support. Religious leaders, scientists, politicians, philosophers, and ethicists argue against the morality and acceptability of human cloning. Nearly all objections hinge, to various degrees, on the definition of human life, beliefs about its sanctity, and the potentially adverse consequences for families and society as a whole.

In an effort to stimulate consideration of and debate about this critical issue, the President's Council on Bioethics examined the principal moral and ethical objections to human cloning in *Human Cloning and Human Dignity: An Ethical Inquiry* (July 2002, http://www.bioethics.gov/reports/cloningreport/fullreport.html). The council's report distinguished between therapeutic and reproductive cloning and outlined key concerns by trying to respond to many as yet unresolved questions about the ethics, morality, and societal consequences of human cloning.

The council determined that the key moral and ethical objections to therapeutic cloning—cloning for biological research—center on the moral status of developing human life. Therapeutic cloning involves the deliberate production, use, and, ultimately, destruction of cloned human embryos. One objection to therapeutic cloning is that cloned embryos produced for research are no different from those that could be used in attempts to create cloned children. Another argument that has been made is that the ends do not justify the means—that research on any human embryo is morally unacceptable, even if this research promises cures for many dreaded diseases. Finally, there are concerns that acceptance of therapeutic cloning will lead society down the "slippery slope" to reproductive cloning, a prospect that is almost universally viewed as unethical and morally unacceptable.

The unacceptability of human reproductive cloning stems from the fact that it challenges the basic nature of human procreation, redefining having children as a form of manufacturing. Human embryos and children may then be viewed as products and commodities rather than as sacred and unique human beings. Furthermore, reproductive cloning might substantially change fundamental issues of human identity and individuality, and allowing parents unprecedented genetic control of their offspring may significantly alter family relationships across generations.

The council concluded that "the right to decide" whether to have a child does not include the right to have a child by any means possible, nor does it include the right to decide the kind of child one is going to have. A societal commitment to freedom does not require use or acceptance of every technological innovation available.

Legislation Aims to Completely Ban Human Cloning

On February 27, 2003, the U.S. House of Representatives voted to outlaw all forms of human cloning. The legislation prohibits the creation of cloned human embryos for medical research as well as the creation of cloned babies. It contains strong sanctions, imposing a maximum penalty of $1 million in civil fines and as many as ten years in jail for violations. The measure did not pass in the Senate, which was closely divided about whether therapeutic cloning should be prohibited along with reproductive cloning. In early February 2003 President Bush issued a policy statement that strongly supported a total ban on cloning. In the Senate two bills were introduced: S. 245 was a complete ban intended to

amend the Public Health Service Act to prohibit all human cloning, and S. 303 was a less sweeping measure that also prohibited cloning but protected stem cell research. S. 245 was referred to the Senate Committee on Health, Education, Labor and Pensions and S. 303 was referred to the Senate Committee on the Judiciary. Neither bill, nor any comparable proposed legislation, has emerged from the Senate committees.

Even though nearly all lawmakers concur that Congress should ban reproductive cloning, many disagree about whether legislation should also ban the creation of cloned human embryos that serve as sources of embryonic stem cells. Many legislators agree with scientists that stem cells derived from cloned human embryos have medical and therapeutic advantages over those derived from conventional embryos or adults. Those who oppose the legislation calling for a total ban assert that the aim of allowing research is to relieve the suffering of people with degenerative diseases. They say that the bill's sponsors are effectively thwarting advances in medical treatment and biomedical innovation.

Supporters of the total ban contend that Congress must send an unambiguous message that cloning research and experimentation will not be tolerated. They consider cloning immoral and unethical, fear unintended consequences of cloning, and feel they speak for the public when they assert that it is not justifiable to create human embryos simply for the purpose of experimenting on them and then destroying them.

The fact that there are only about twenty available stem cell lines prompted the introduction of bills during the spring of 2004 that would require funding for human embryonic stem cell research, despite the president's 2001 policy. On April 28, 2004, more than 200 members of the House sent a letter to the president arguing in favor of an expansion of existing policy. Fifty-eight senators sent a similar letter on June 4, 2004. Pleas from patient advocacy groups—along with the death of the former president Ronald Reagan from Alzheimer's disease on June 5, 2004, and Nancy Reagan's appeals to expand the policy—focused considerable media attention on the issue during the summer of 2004, but no legislation was passed that year.

On May 24, 2005, the House passed H.R. 810, the Stem Cell Research Enhancement Act of 2005, which would have permitted federal funding for embryonic stem cell research on cells "derived from human embryos that have been donated from in vitro fertilization clinics, were created for the purposes of fertility treatment, and were in excess of the clinical need of the individuals seeking such treatment." The Senate passed the bill on July 18, 2006, and the following day President Bush vetoed the bill.

TABLE 8.1

State human cloning laws, April 2006

State	Statute citation	Summary	Prohibits reproductive cloning	Prohibits therapeutic cloning	Expiration
Arizona	HB 2221 (2005)	Bans the use of public monies for reproductive or therapeutic cloning	Prohibits use of public monies	Prohibits use of public monies	
Arkansas	§20–16–1001 to 1004	Prohibits therapeutic and reproductive cloning; may not ship, transfer or receive the product of human cloning; human cloning is punishable as a class C felony and by a fine of not less than $250,000 or twice the amount of pecuniary gain that is received by the person or entity, which ever is greater	yes	yes	
California	Business And Professions §16004–5 Health & Safety §24185, §24187, §24189, §12115–7	Prohibits reproductive cloning; permits cloning for research; provides for the revocation of licenses issued to businesses for violations relating to human cloning; prohibits the purchase or sale of ovum, zygote, embryo, or fetus for the purpose of cloning human beings; establishes civil penalties	yes	no	
Connecticut	2005 SB 934	Prohibits reproductive cloning, permits cloning for research; punishable by not more than one hundred thousand dollars or imprisonment for not more than ten years, or both	yes	no	
Indiana	2005 Senate Enrolled Act No. 268	Prohibits reproductive and therapeutic cloning; allows for the revocation of a hospital's license involved in cloning; specifies that public funds may not be used for cloning; prohibits the sale of a human ovum, zygote, embryo or fetus	yes	yes	
Iowa	707B.1 to 4	Prohibits human cloning for any purpose; prohibits transfer or receipt of a cloned human embryo for any purpose, or of any oocyte, human embryo, fetus, or human somatic cell, for the purpose of human cloning; human cloning punishable as class C felony; shipping or receiving punishable as aggravated misdemeanor; if violation of the law results in pecuniary gain, then the individual is liable for twice the amount of gross gain; a violation is grounds for revoking licensure or denying or revoking certification for a trade or occupation	yes	yes	
Maryland	2006 SB 144	Prohibits reproductive cloning; prohibits donation of oocytes for state-funded stem cell research but specifies that the law should not be construed to prohibit therapeutic cloning; prohibits purchase, sale, transfer or obtaining unused material created for in vitro fertilization that is donated to research; prohibits giving valuable consideration to another person to encourage the creation of in vitro fertilization materials solely for the purpose of research; punishable by up to three years in prison; a maximum fine of $50,000 or both	yes	no	
Massachusetts	2005 SB 2039	Prohibits reproductive cloning; permits cloning for research; prohibits a person from purchasing, selling, transferring, or obtaining a human embryonic, gametic or cadaveric tissue for reproductive cloning; punishable by imprisonment in jail or correctional facility for not less than five years or more than ten years or by or by imprisonment in state prison for not more than ten years or by a fine of up to one million dollars; in addition a person who performs reproductive cloning and derives financial profit may be ordered to pay profits to commonwealth	yes	no	
Michigan	§§333.2687–2688, §§333.16274–16275, 333.20197, 333.26401–26403, 750.430a	Prohibits human cloning for any purpose and prohibits the use of state funds for human cloning; establishes civil and criminal penalties	yes	yes	
Missouri	§1.217	Bans use of state funds for human cloning research which seeks to develop embryos into newborn child	Prohibits the use of state funds	no	
New Jersey	§2C:11A-1, §26:2Z-2	Permits cloning for research; prohibits reproductive cloning, which is punishable as a crime in the first degree; prohibits sale purchase, but not donation, or embryonic or fetal tissue, which is punishable as a crime in the third degree and a fine of up to $50,000	yes	no	

State Human Cloning Laws

As of 2006 fifteen states had enacted legislation that addresses human cloning. (See Table 8.1.) California was the first state to ban reproductive cloning in 1997. Since then, twelve other states—Arkansas, Connecticut, Indiana, Iowa, Maryland, Massachusetts, Michigan, Rhode Island, New Jersey, North Dakota, South Dakota, and Virginia— have passed laws prohibiting reproductive cloning. Arizona's and Missouri's legislation addresses the use of public funds for cloning, and Maryland's prohibits

TABLE 8.1

State human cloning laws, April 2006 [CONTINUED]

State	Statute citation	Summary	Prohibits reproductive cloning	Prohibits therapeutic cloning	Expiration
North Dakota	§12.1–39	Prohibits reproductive and therapeutic cloning; transfer or receipt of the product of human cloning; transfer or receipt, in whole or in part, any oocyte, human embryo, human fetus, or human somatic cell, for the purpose of human cloning; cloning or attempt to clone punishable as a class C felony; shipping or receiving violations punishable as class A misdemeanor	yes	yes	
Rhode Island	§23–16.4–1 to 4–4	Prohibits human cloning for the purpose of initiating a pregnancy; for a corporation, firm, clinic, hospital, laboratory, or research facility, punishable by a civil penalty punishable by fine of not more than $1,000,000, or in the event of pecuniary gain, twice the amount of gross gain, whichever is greater; for an individual or an employee of the firm, clinic, hospital, laboratory, or research facility acting without the authorization of the firm, clinic, hospital, or research facility, punishable by a civil penalty punishable by fine of not more than $250,000, or in the event of pecuniary gain, twice the amount of gross gain, whichever is greater	yes	no	July 7, A2010
South Dakota	§34–14–27	Prohibits reproductive and therapeutic cloning; transfer or receipt of the product of human cloning; transfer or receipt, in whole or in part, any oocyte human embryo, human fetus, or human somatic cell, for the purpose of human cloning; cloning or attempt to clone is punishable as a felony and a civil penalty of gross gain, or any intermediate	yes	yes	
Virginia	§32.1–162.32–2	Prohibits reproductive cloning; may prohibit therapeutic cloning but it is unclear because human being is not defined in the definition of human cloning; human cloning defined as the creation of or attempt to create a human being by transferring the nucleus from a human cell from whatever source into an oocyte from which the nucleus has been removed; also prohibits the implantation or attempted implantation of the product of somatic cell nuclear transfer into an uterine environment so as to initiate a pregnancy; the possession of the product of human cloning; and the shipping or receiving of the product of a somatic cell nuclear transfer in commerce for the purpose of implantation of such product into an uterine environment so as to initiate a pregnancy. The law establishes civil penalty not to exceed $50,000 for each incident.	yes	unclear	

SOURCE: "State Human Cloning Laws," National Conference of State Legislatures, April 18, 2006, http://www.ncsl.org/programs/health/Genetics/rt-shcl.htm (accessed October 30, 2006)

the use of state stem cell research funds for reproductive cloning and possibly therapeutic cloning, depending the interpretation of the statute. Louisiana also enacted legislation that prohibited reproductive cloning, but the law expired in July 2003. The laws of Arkansas, Indiana, Iowa, Michigan, North Dakota, and South Dakota also prohibit therapeutic cloning. Virginia's legislation may be interpreted as a complete ban on human cloning; however, it is unclear because the law does not define the term *human being*, which is used in the definition of human cloning. Rhode Island's law does not prohibit cloning for research, and California's and New Jersey's laws specifically permit cloning for the purpose of research.

California Leads the Way

In 2002 the California state legislature passed a law encouraging therapeutic cloning. Even though there were no provisions for funds in the law, the move was interpreted as support for the research. In 2004 stem cell research advocates offered voters a sweeping ballot measure—Proposition 71—to make public funding available to support stem cell research and therapeutic cloning. Proposition 71 was championed by Robert Klein, a wealthy real estate developer and father of a child with diabetes who might benefit from the research. It also received considerable financial support from the Microsoft founder Bill Gates to finance campaign advertising and lobbying.

On November 2, 2004, Californians approved Proposition 71, a ballot measure with the potential to make the state a leader in human embryonic stem cell research. Proposition 71 enabled the state to establish its own research institute—the California Institute for Regenerative Medicine. The proposition prohibits reproductive cloning but funds human cloning projects designed to create stem cells and allocates $3 billion over ten years in research funds. Those supporting the legislation hoped that stem cell research would become the biggest, most important, and most profitable medical advancement of the twenty-first century. The legislation's supporters intended to use the funds to attract top researchers to

FIGURE 8.5

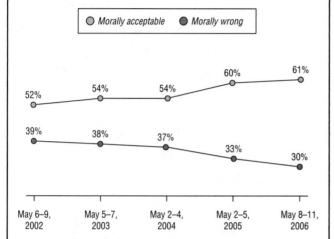

Public opinion on the moral acceptability of embryonic stem cell research, 2002–06

NEXT, I'M GOING TO READ YOU A LIST OF ISSUES. REGARDLESS OF WHETHER YOU THINK IT SHOULD BE LEGAL, FOR EACH ONE, PLEASE TELL ME WHETHER YOU PERSONALLY BELIEVE THAT IN GENERAL IT IS MORALLY ACCEPTABLE OR MORALLY WRONG. HOW ABOUT—MEDICAL RESEARCH USING STEM CELLS OBTAINED FROM HUMAN EMBRYOS?

● Morally acceptable   ● Morally wrong

52%   54%   54%   60%   61%

39%   38%   37%   33%   30%

May 6–9, 2002   May 5–7, 2003   May 2–4, 2004   May 2–5, 2005   May 8–11, 2006

SOURCE: "Next, I'm going to read you a list of issues. Regardless of whether you think it should be legal, for each one, please tell me whether you personally believe that in general it is morally acceptable or morally wrong. How about—medical research using stem cells obtained from human embryos?" in *Stem Cell Research*, The Gallup Organization, 2006, http://www.galluppoll.com:/content/Default.aspx?ci=21676&pg=1&t=K8cSWpBM9.VI6kGw%2f-FjYmf9f9v1wvl6KJ XHZitkW8XKNmRQQR-rNsIPiOO1CYyUAUhNRGXzigw8p72Otcyo Qrljyt-4if%2fZCTPAtbbQ%2fnW0mrEkB-PDjFPelHT35F9PEhtjy2iG 6ECYqHVBPdIoPGAE4CnSTBFnZ.Jp0RCZK2Ua (accessed November 1, 2006). Copyright © 2006 by The Gallup Organization. Reproduced by permission of The Gallup Organization.

FIGURE 8.6

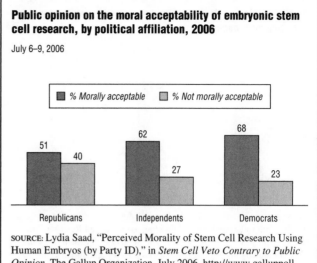

Public opinion on the moral acceptability of embryonic stem cell research, by political affiliation, 2006

July 6–9, 2006

■ % Morally acceptable   □ % Not morally acceptable

51   40   62   27   68   23

Republicans   Independents   Democrats

SOURCE: Lydia Saad, "Perceived Morality of Stem Cell Research Using Human Embryos (by Party ID)," in *Stem Cell Veto Contrary to Public Opinion*, The Gallup Organization, July 2006, http://www.galluppoll.com/content/default.aspx?ci=23827&pg=1 (accessed November 1, 2006). Copyright © 2006 by The Gallup Organization. Reproduced by permission of The Gallup Organization.

the state, making California the epicenter of groundbreaking, lifesaving, and potentially lucrative medical research.

Nicholas Wade reports in "Plans Unveiled for State-Financed Stem Cell Work in California" (*New York Times*, October 5, 2006) that in October 2006 the California Institute for Regenerative Medicine released its ten-year plan for spending the $3 billion allocated to it. The institute said it will spend $823 million on basic stem cell research, $899 million on applied or preclinical research, and $656 million to advance new treatments through clinical trials. An additional $273 million will enable universities to construct laboratories in which none of the equipment has been purchased with federal funds to ensure that the researchers are not violating the rules that restrict federal money to conduct stem cell research.

Public Opinions about Stem Cell Research and Cloning

According to Gallup poll data, more than 60% of Americans believe using stem cells derived from human embryos in medical research is morally acceptable. Figure 8.5 reveals that the percentage of Americans that considers stem cell research morally acceptable had increased from 52% in 2002 to 61% in 2006.

The percentage of Americans that deems stem cell research morally acceptable varies by political affiliation, with support highest among Democrats (68%) and Independents (62%), compared with Republicans (51%). (See Figure 8.6.) According to Lydia Saad in *Stem Cell Veto Contrary to Public Opinion* (Gallup Poll, July 20, 2006), support also varies by educational attainment—three-quarters (77%) of those with postgraduate degrees consider this research acceptable, compared with 45% of people who had attained a high school education or less.

The Gallup poll also found that most Americans (58%) disapproved of President Bush's July 2006 veto of a bill that would have expanded federal funding for embryonic stem cell research. (See Figure 8.7.) However, Saad notes that just 11% of Americans favor unrestricted government funding of embryonic stem cell research and another 42% support easing current restrictions. Nearly one-quarter (24%) approve of the current funding restrictions and 19% oppose any government funding of this research.

Even though Americans continue to feel that it is morally unacceptable to clone humans, public support for cloning animals increased slightly from 31% in 2001 to 35% in 2005. (See Figure 8.8.) Furthermore, unlike stem cell research, which is favored by more Democrats than Republicans; more Republicans (31%) than Democrats (28%) consider cloning animals morally acceptable. (See Figure 8.9 and Figure 8.10.)

FIGURE 8.7

Public opinion on President Bush's veto of expanded funding for stem cell research, 2006

July 21–23, 2006

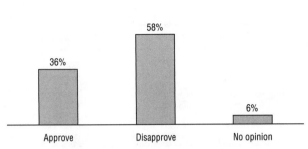

SOURCE: Joseph Carroll, "Public Approval for President Bush's Decision to Veto Bill to Expand Federal Funding for Stem Cell Research," in *Public Opposes President Bush's Veto on Stem Cell Research Funding*, The Gallup Organization, July 2006, http://www.galluppoll.com/content/?ci=23902&pg=1 (accessed November 1, 2006). Copyright © 2006 by The Gallup Organization. Reproduced by permission of The Gallup Organization.

FIGURE 8.8

Public opinion on the moral acceptability of cloning animals and humans, 2001–05

[Numbers shown in percentages]

NEXT, I'M GOING TO READ YOU A LIST OF ISSUES. REGARDLESS OF WHETHER OR NOT YOU THINK IT SHOULD BE LEGAL, FOR EACH ONE, PLEASE TELL ME WHETHER YOU PERSONALLY BELIEVE THAT IN GENERAL IT IS MORALLY ACCEPTABLE OR MORALLY WRONG.

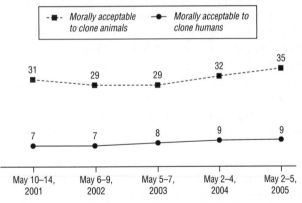

SOURCE: "Next, I'm going to read you a list of issues. Regardless of whether you think it should be legal, for each one, please tell me whether you personally believe that in general it is morally acceptable or morally wrong," in *Cloning*, The Gallup Organization, 2005, http://www.galluppoll.com/content/?ci=6028&pg=1 (accessed November 1, 2006). Copyright © 2006 by The Gallup Organization. Reproduced by permission of The Gallup Organization.

FIGURE 8.9

Moral acceptability of cloning and other issues among Democrats, 2006

[Numbers shown in percentages]

May 8–11, 2006

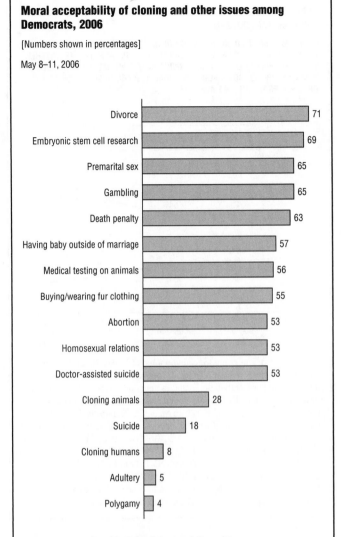

SOURCE: Joseph Carroll, "Moral Acceptability of Issues among Democrats (Including Leaners)," in *Republicans, Democrats Differ on What Is Morally Acceptable*, The Gallup Organization, May 2006, http://www.galluppoll.com/content/?ci=22915&pg=1 (accessed November 1, 2006). Copyright © 2006 by The Gallup Organization. Reproduced by permission of The Gallup Organization.

FIGURE 8.10

Moral acceptability of cloning and other issues among Republicans, 2006

[Numbers shown in percentages]

May 8–11, 2006

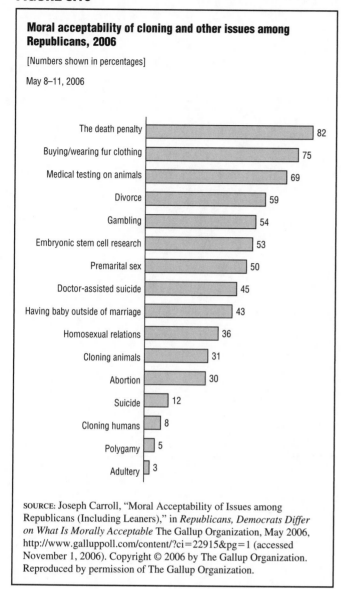

SOURCE: Joseph Carroll, "Moral Acceptability of Issues among Republicans (Including Leaners)," in *Republicans, Democrats Differ on What Is Morally Acceptable* The Gallup Organization, May 2006, http://www.galluppoll.com/content/?ci=22915&pg=1 (accessed November 1, 2006). Copyright © 2006 by The Gallup Organization. Reproduced by permission of The Gallup Organization.

CHAPTER 9
GENETIC ENGINEERING AND BIOTECHNOLOGY

Around the world scientists are working to develop new varieties of crops that can resist pests, use less water, and generally thrive in less than optimal growing conditions. Hand in hand, with scientific research, countries must adopt policies that will allow their farmers to take advantage of new products being developed through research. Government policies should encourage the safe use of new technology, not cause farmers and consumers to fear it.

—Ann M. Veneman, U.S. Secretary of Agriculture, October 2002

The dawn of the new millennium saw explosive advances in biotechnology. Technological breakthroughs offered scientists and physicians unprecedented opportunities to develop previously inconceivable solutions to pressing problems in agriculture, environmental science, and medicine. Simultaneously, researchers, politicians, ethicists, theologians, and the public were challenged to assess, analyze, and determine the feasibility of using new biotechnology in view of the opportunities, possibilities, risks, benefits, and diverse viewpoints about the safe, effective, and ethical applications of genetic research.

This chapter describes several examples of existing and proposed applications of genetic research and biotechnology, including uses that address such pressing problems as environmental pollution and world hunger as well as the role of genetic engineering in developing lifesaving medical therapeutics. It also considers industry and consumer viewpoints as well as recommendations scientists, policy makers, and ethicists have made regarding the wise, judicious, and equitable use of these technologies.

AGRICULTURAL APPLICATIONS OF GENETIC ENGINEERING

Genetically modified (GM) or transgenic crops (sometimes also called genetically engineered [GE] crops) contain one or more genes that have been artificially inserted instead of received through pollination (fertilization by the transfer of pollen from an anther to a stigma of a plant). The inserted gene sequence, called the transgene, may be introduced to produce different results—either to overexpress or silence (direct a gene not to synthesize a specific protein) an existing plant gene, and it may come from another unrelated plant or from a completely different species. Transgenics is the science of inserting a foreign gene into an organism's genome. The ultimate product of this technology is a transgenic organism.

For example, the transgenic corn that produces its own insecticide contains a gene from a bacterium, and Macintosh apples with a gene from a moth that encodes an antimicrobial protein are resistant to fire blight, a bacterial infection. Although all crops have been genetically modified from their original wild state by domestication, selection, and controlled breeding over long periods of time, the terms *transgenic crops* and *GM crops* usually refer to plants with transgenes (inserted gene sequences).

Genes are introduced into a crop plant to make it as useful and productive as possible by acting to protect the plant, improve the harvest, or enable the plant to perform a new function or acquire a new trait. Specific objectives of genetically modifying a plant include increasing its yield, improving its quality, or enhancing its resistance to pests or disease and its tolerance for heat, cold, or drought. Some of the GM traits that have been introduced into food crops are enhanced flavor, slowed ripening, reduced reliance on fertilizer, self-generating insecticide, and added nutrients. Examples of transgenic food crops include frost-resistant strawberries and tomatoes; slow-ripening bananas, melons, and pineapples; and insect-resistant corn.

Transgenic technology enables plant breeders to bring together in one plant useful genes from a wide range of living sources, not just from within the crop species or from closely related plants. It provides reliable means for identifying and isolating genes that control specific characteristics in one kind of organism and enables researchers to move copies of these genes into another organism that will then develop the chosen characteristics. This technology gives plant breeders the ability to generate more useful and productive crop varieties containing new combinations of genes, and it significantly expands the range of trait manipulation and enhancements well beyond the limitations of traditional cross-pollination and selection techniques.

Although genetic modification of plants generates the same types of changes produced by conventional agricultural techniques, because it precisely alters a single gene, the results are often more rapid and more complete. Traditional breeding techniques may require an entire generation or more to introduce or remove a single gene, and using conventional methods for breeding a polygenic trait into crops with multiyear generations could take several decades.

Creating Transgenic Crops

The first step in creating a transgenic plant is locating genes with the traits that growers, marketers, and consumers consider important. These are usually genes that increase productivity and yield and improve resistance to environmental stresses such as frost, heat, salt, and insects. Identifying the gene associated with a specific trait is necessary but not sufficient; researchers must determine how the gene is regulated, its other influences on the plant, and its interactions with other genes to express or silence various traits. Researchers must then isolate and clone the gene to have sufficient quantities to modify. Establishing the genomic sequence, called plant genomics, and the functions of genes of the most important crops is a priority of public- and private-sector plant genomic research projects.

Currently, most genes introduced into plants come from bacteria; however, increasing understanding of plant genomics is anticipated to permit greater use of plant-derived genes to genetically engineer crops. In 2000 the first entire plant genome *Arabidopsis thaliana* was sequenced, which provided researchers with new insight into the genes that control specific traits in many other agricultural plants. There are several approaches to introducing genes into plant cells: vector- or carrier-mediated transformation, particle-mediated transformation, and direct deoxyribonucleic acid (DNA) insertion.

Vector-mediated transformation involves infecting plant cells with a virus or bacterium that during the process of infection inserts foreign DNA into the plant cell. The most convenient is through the soil bacterium *Agrobacterium tumefaciens*, which infects tomatoes, potatoes, cotton, and soybeans. This bacterium attacks cells by inserting its own DNA. When genes are added to the bacterium, they are transferred to the plant cell along with the other DNA.

FIGURE 9.1

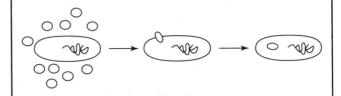

A bacterial cell incorporating DNA in a process known as transformation

SOURCE: Adapted from "Step 3. Transformation," in *Joint Genome Institute: Education—How Sequencing Is Done*, U.S. Department of Energy Office of Science, Joint Genome Institute, September 2004, http://www.jgi.doe.gov/education/how/how_3.html (accessed December 15, 2006)

Particle-mediated transformation involves gene transfer using a special particle tool known as a gene gun, which shoots tiny metal particles that contain DNA into the cell.

To perform direct DNA insertion or electroporation, cells are immersed in the DNA and electrically shocked to stimulate DNA uptake. The cell wall then opens for less than a second, allowing DNA to seep into the cell. (See Figure 9.1.) Following gene insertion, the cell incorporates the foreign DNA into its own chromosomes and undergoes normal cell division. The new cells ultimately form the organs and tissues of the "regenerated" plant. To ensure the systematic sequence of these steps, other genes may be added along with the gene associated with the desired trait. These helper genes are called promoters. They encourage growth of cells that have integrated the inserted DNA, provide resistance to stresses (such as toxins present in the medium used to grow the cells), and may help regulate the functions of the gene linked to the desired trait.

To be certain that the new genes are in the organism, marker genes are sometimes inserted along with the gene for the desired trait. One common marker gene confers resistance to the antibiotic kanamycin. When this gene is used as a marker, investigators are able to confirm that the transfer was successful when the organism resists the antibiotic. The ultimate success of gene insertion is measured by whether the inserted gene functions properly by expressing, amplifying, or silencing the desired trait.

Are GM Crops Helpful or Harmful?

In *Genes Are Gems: Reporting Agri-Biotechnology* (December 2006, http://www.icrisat.org/Publications/Genes_Gems.htm), the International Crops Research Institute for the Semi-Arid Tropics (ICRISAT), an international nonprofit and nonpartisan organization for science-based agricultural development, notes that genetic modification of crops has proven to be the most rapidly adopted technology in the world. Figure 9.2 shows the

FIGURE 9.2

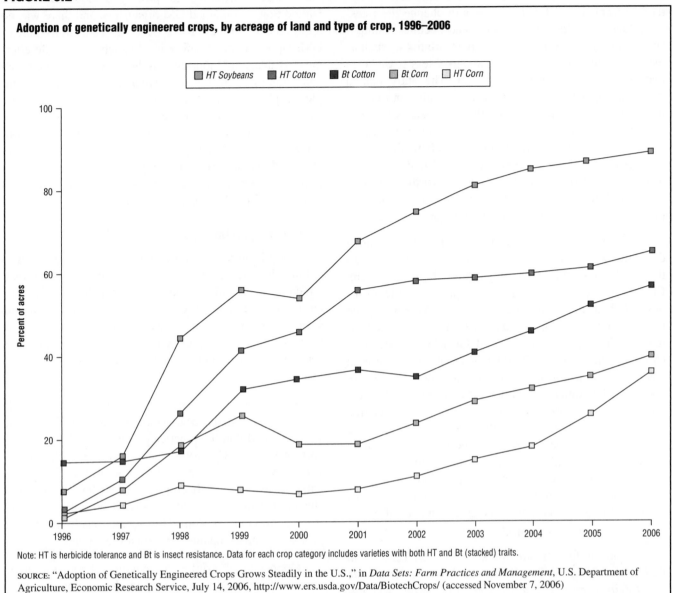

Adoption of genetically engineered crops, by acreage of land and type of crop, 1996–2006

Note: HT is herbicide tolerance and Bt is insect resistance. Data for each crop category includes varieties with both HT and Bt (stacked) traits.

SOURCE: "Adoption of Genetically Engineered Crops Grows Steadily in the U.S.," in *Data Sets: Farm Practices and Management*, U.S. Department of Agriculture, Economic Research Service, July 14, 2006, http://www.ers.usda.gov/Data/BiotechCrops/ (accessed November 7, 2006)

widespread adoption of GM crops by U.S. farmers since their introduction in 1996. ICRISAT contends that transgenic crops benefit developing countries by enabling greater use of crop area, increasing the variety of crops that may be grown, affording better protection of crops from disease and pests, and improving harvest yields to deliver more food and nutrition to people in need. ICRISAT also cites environmental benefits of transgenic organisms. These include a 30% to 40% reduction in the use of herbicides and as much as an 80% reduction in the use of insecticides, as well as a reduction in environmental pollution and harmful emissions resulting from the production of chemical pesticides.

ICRISAT also commends the use of transgenic organisms, specifically transgenic microbes (microorganisms) for purposes of bioremediation (environmental cleanup). Researchers have harnessed genes that code for proteins

that naturally degrade toxic wastes such as chlorinated pesticides, naphthalene, toluene, and some hydrocarbons. Efforts are under way to combine genes from several microbes to create a single, multipurpose supermicrobe that is capable of effectively combating several contaminants.

Opposition to GM crops takes several forms. Bioethicists who contend that freedom of choice is a central tenet of ethical science oppose what they deem to be interference with other forms of life. Environmentalists argue that transgenic technology poses the risk of altering delicately balanced ecosystems—biological communities and their environments—and causing unintended harm to other organisms. They are concerned that transgenic crops will replace traditional crop varieties, especially in developing countries, causing loss of biological diversity.

In "Turning Genetically Engineered Trees into Toxic Avengers" (*New York Times*, August 3, 2004), Hillary

Resner reports on one controversial example: the relatively recent use of GE forest trees to change the trees' reproductive cycles, growth rates, and chemical composition so they can resist disease and absorb toxins such as mercury from soil and convert it into a less toxic form that is safe for release into the air. The aptly named "toxic-avenger trees" remove heavy metals from contaminated soils in places where other approaches to environmental cleanup are costly and labor intensive. Environmentalists are concerned, however, about the use of GE trees because they are not convinced that relocating heavy metals from the soil to the air is worth the risk of the altered genes migrating via the tree pollen to natural populations, potentially damaging existing ecosystems.

Pests may develop resistance to transgenics in much the same way certain bacteria have become resistant to the antibiotics that once effectively eradicated them. Critics also decry the infiltration of transgenic crops beyond their intended areas and fear inadvertent gene transfer to species not targeted for transgenics. Rex Dalton, in "Transgenic Corn Found Growing in Mexico" (*Nature*, September 27, 2001), notes that transgenic corn has been found in a remote mountain region of Mexico, and K. S. Jayaraman, in "Illicit GM Cotton Sparks Corporate Fury" (*Nature*, October 11, 2001), reports that transgenic cotton has been discovered in India. One way unintended gene exchange between plants may occur is through pollen. Recommendations for preventing unintended gene exchange include creation of transgenic plants that do not produce pollen or of pollen that does not contain introduced genes and establishment of buffer zones around fields of transgenic crops.

According to Graham Brookes and Peter Barfoot, in *GM Crops: The First Ten Years—Global Socio-economic and Environmental Impacts* (2006, http://www.isaaa.org/ Resources/Publications/briefs/36/download/isaaa-brief-36 -2006.pdf), transgenic crops are big business. Brookes and Barfoot estimate that worldwide, the net economic benefit to biotech crop farmers in 2005 was $5.6 billion, $6.15 billion in 2006, and was projected to approach $7 billion in 2007. Opponents fear consequences such as economic concentration—the potential for companies that grow transgenic crops to drive out smaller farmers and create monopolies. In 2006 the techniques to genetically modify seeds, as well as the seeds themselves, were held by a few multinational corporations. Related issues are patent infringement and intellectual property rights for transgenic crops and absence of regulatory oversight. Patents may increase the price of seeds and effectively exclude small farmers from growing their crops.

Health risks also concern those who object to widespread acceptance of transgenic crops. They call for the labeling of GM food to alert consumers that they are purchasing foods that contain GM organisms. As of 2006 more than half of all processed foods sold in the United States contained GM organisms, and there was no requirement that these foods be identified as transgenic or GM. Opponents cite safety issues such as possible allergies to transgenic foods and products because some transgenes may pose health risks when consumed. For example, a plan to insert a Brazil nut protein gene into soybeans was halted when early tests indicated that people allergic to nuts suffered reactions when they consumed the modified soy products. Critics also fear that there will be unforeseen and potentially harmful long-term adverse health consequences resulting from the consumption of foods containing foreign genes.

Terminator Technology

One of the most controversial developments in agricultural bioengineering is called terminator technology, which is designed to genetically switch off a plant's ability to germinate a second time. Traditionally, farmers save seeds for the next harvest; however, the use of terminator technology effectively prevents this practice, forcing them to purchase a fresh supply of seeds each year.

The advocates of terminator technology are generally corporations and the organizations that represent them. They contend that the practice protects corporations from corrupt farmers. Controlling seed germination helps prevent growers from pirating the corporations' licensed or patented technology. If crops remained fertile, there is a chance that farmers could use any saved transgenic seed from a previous season. This would result in reduced profits for the companies that own the patents.

Opponents of terminator technology believe it threatens the livelihood of farmers in developing countries such as India, where many poorer farmers have been unable to compete and some have been forced out of business. Opponents considered it a victory when Monsanto, a major investor in this technology, decided not to market terminator technology. In *Terminator Technology—Five Years Later* (May–June 2003, http:// www.cbdcprogram.org/final/issues/termcom79eng.pdf), the ETC Group, an organization that focuses on conservation and sustainable advancement of cultural and ecological diversity and human rights, notes that besides Monsanto, terminator patents are held by Delta and Pine Land, the U.S. Department of Agriculture, Syngenta, DuPont, and BASF, as well as the universities Purdue, Iowa State, and Cornell. According to the Pesticide Action Network North America, in "Corporate Greed + Destructive Technology = Increased Risk of World Hunger: Terminator Seed Moratorium at Risk" (January 10, 2006, http://www.panna.org/resources/panups/panup_ 20060110.dv.html), Syngenta owns or has applied for eleven terminator patents—more than any other company—but the company has stated publicly that it will not commercialize the trait.

Still, even without terminator technology, under patent laws in Canada, the United States, and many other industrialized nations, it is illegal for farmers to reuse patented seed or to grow Monsanto's GM seed without signing a licensing agreement. This has the same effect on poor farmers as terminator technology; it renders them unable to compete. Jeffrey L. Fox reports in "Canadian Farmer Found Guilty of Monsanto Canola Patent Infringement" (*Nature Biotechnology*, May 2001) that a Canadian farmer was found guilty of growing patented seeds, even though he did it inadvertently. Pollen from the patented canola seeds at a nearby farm had pollinated his plants, and he was ordered to pay Monsanto for licensing and profit from the seeds.

In *Monsanto vs. U.S. Farmers* (January 13, 2005, http://www.centerforfoodsafety.org/pubs/CFSMOnsantovs FarmerReport1.13.05.pdf), the Center for Food Safety (CFS), a nonprofit public interest and environmental advocacy organization, reviewed Monsanto's legal actions against U.S. farmers. The report finds that Monsanto engaged in investigations of farmers, out-of-court settlements, and litigation against farmers allegedly in breach of contract or engaged in patent infringement. The report documents 90 Monsanto lawsuits in 25 states that involve 147 farmers and 39 small businesses or farm companies. Monsanto has an annual budget of $10 million and seventy-five employees exclusively devoted to investigating and taking action against farmers.

According to the CFS, by 2005 the largest judgment in favor of Monsanto was more than $3 million, and the total recorded judgments granted to Monsanto was more than $15 million. Farmers paid a mean of $412,259 for cases with recorded judgments. Some farmers were even forced to pay Monsanto's costs while they were under investigation.

In the CFS press release "Monsanto Assault on U.S. Farmers Detailed in New Report" (January 13, 2005, http://www.centerforfoodsafety.org:80/press_release1.13 .05.cfm), Andrew Kimbrell, the executive director of the CFS, asserts that "these lawsuits and settlements are nothing less than corporate extortion of American farmers. Monsanto is polluting American farms with its genetically engineered crops, not properly informing farmers about these altered seeds, and then profiting from its own irresponsibility and negligence by suing innocent farmers. We are committed to stopping this corporate persecution of our farmers in its tracks."

On September 29, 2006, the Public Patent Foundation (http://www.pubpat.org/monsantofiled.htm), a not-for-profit legal services organization that represents the public's interests against the harms caused by the patent

system, filed requests with the U.S. Patent and Trademark Office to revoke four patents owned by Monsanto "that the agricultural giant is using to harass, intimidate, sue—and in some cases literally bankrupt—American farmers."

Exorcist Technology

In 2004 the biotechnology industry introduced what it calls exorcist technology to some GE crops. This new technology introduces chemical inducers that shed their foreign DNA before they are harvested. The industry sees this technology as an effective way to counter anti-GE critics because the harvested crops will not contain foreign DNA. However, detractors assert that the intent of exorcist technology is to shift the responsibility from the biotechnology industry to farmers and society. If gene flow poses a problem, farmers will have to use chemical inducers to remove the offensive transgenes.

U.S. BIOTECHNOLOGY REGULATORY SYSTEM

The U.S. government operates a rigorous, coordinated regulatory process for determining the safety of agricultural products of modern biotechnology. The process ensures that all biotechnology products that are commercially grown, processed, sold, and consumed are as safe as their conventional counterparts. In *A Description of the U.S. Food Safety System* (March 3, 2000, http://www.fsis .usda.gov/OA/codex/system.htm), the U.S. Department of Agriculture (USDA) notes that the government regulatory system is "transparent, predictable, open to public comment, and based on sound science." Within the USDA the agencies responsible for regulation are the Animal and Plant Health Inspection Service (APHIS) and the Food and Drug Administration (FDA). The Environmental Protection Agency (EPA)—which is not a part of the USDA—also plays a role in regulating biotechnology. Policies, processes, and regulations are continuously reviewed, evaluated, and, when necessary, revised to meet the challenges of this evolving technology.

APHIS oversees U.S. agriculture, protecting against pests and diseases. It is the lead agency regulating the safe field-testing of biotechnology-derived new plant varieties and certain microorganisms. As such, APHIS grants approval and licenses for veterinary biological substances including animal vaccines that may be the products of biotechnology.

The EPA approves new herbicidal and pesticidal substances. It issues permits for testing herbicides and biotechnology-derived plants containing new pesticides. When the EPA makes a determination about whether to register a new pesticide, it considers human safety, environmental impact, its effectiveness on the target pest, and any consequences for other, nontarget species. The EPA establishes and enforces the guidelines that ensure safe use of GE products classified as pesticides.

The FDA ensures that foods derived from new bio-engineered plant varieties are safe and nutritious—they must meet or exceed the same high standards of safety applied to any food product. The FDA is also responsible for issuing and enforcing regulations to ensure that all food and feed labels, including those related to biotechnology, are truthful and do not mislead consumers. Table 9.1 lists genes, gene fragments, and GM products (with their intended effects) that were submitted to the FDA from 2002 to mid-2005.

Institute of Medicine Report

In July 2004 the Institute of Medicine (IOM) of the National Academy of Sciences, an organization created by the U.S. Congress in 1863 to advise the government about scientific and technical matters, published *The Safety of Genetically Engineered Foods: Approaches to Assessing Unintended Health Effects* (http://www.nap.edu/books/0309092094/html/). The USDA, the FDA, and the EPA commissioned the National Academy of Sciences to assess the potential for adverse health effects from GE foods compared with foods altered in other ways, and to provide guidance on how to identify and evaluate the likelihood of those effects. Even though adverse health effects from genetic engineering have not been detected in the human population, the technique is relatively new and concerns about its safety remain.

The IOM report urged federal agencies to assess the safety of genetically altered foods on a case-by-case basis to determine whether unintended changes in their composition have the potential to adversely affect human health and called for greater scrutiny of foods containing new compounds or unusual amounts of naturally occurring substances.

The report presented a framework to guide federal agencies in selecting the course and intensity of safety assessment. A new GM food whose composition is similar to a commonly used conventional version may warrant little or no additional safety evaluation. If, however, an unknown substance has been detected in a food, more detailed analyses should be conducted to determine whether an allergen or toxin may be present. Similarly, foods with nutrient levels that fall outside the normal range should be assessed for their potential impact on consumers' diets and health.

The IOM was also charged with examining the safety of foods from cloned animals. The report recommended that the safety evaluation of these foods should focus on the product itself rather than on the process used to create it, and advised that the evaluations compare foods from cloned animals with comparable food products from noncloned animals. Although there is no evidence that foods from cloned animals pose an increased risk to consumers, the report cautioned that cloned animals engineered to produce pharmaceuticals should not be permitted to enter the food chain.

Criticisms of the U.S. Regulatory Approach

Opponents of GM foods do not believe there are sufficient government regulations in place to control U.S. production and distribution of these foods. They argue that there has not been enough research or long-term experience with these foods, and as a result the health consequences of growing and eating such foods as well as the environmental impact are not yet known. Proponents claim the benefits of transgenic foods—improved flavor, increased nutritional value, longer shelf life, and greater yields—surpass any potential risks. They also discount the health risks, observing that nearly half the soybean crop and a quarter of all corn grown in the United States consists of transgenic varieties, meaning that Americans have been consuming transgenic food products for years and, as of early 2007, there have been no reports of adverse health effects as a result.

Since the late 1990s there have been protests staged to oppose the widespread use and consumption of GM foods as well as attacks on facilities conducting research on transgenic crops and companies marketing GM products. Protesters have dubbed the transgenic crops "Frankenfoods," likening them to Frankenstein's monster, a man-made freak created through science. Greenpeace, an organization that opposes the creation of GM foods, and other activists staged a historic protest at the World Trade Organization (WTO) meeting in Seattle on November 30, 1999. Greenpeace also hung an anti-GM banner on Kellogg's Cereal City museum in Michigan in 2000. It hopes that such actions will move U.S. lawmakers to require labeling of transgenic foods. Greenpeace is inspired by the example of European consumers, who demanded and have been granted product labeling that enables them to choose whether to purchase and consume GM foods. During 2006 Greenpeace organized protests in cornfields in Spain, the Philippines, and Mexico to protect native corn crops from contamination from genetically engineered strains. The organization also requested an inquiry into the death of cattle in India that had grazed on GM crops ("Greenpeace Opposes GM Crops," Tribune News Service, http://www.tribuneindia.com/2006/20060614/nation.htm#15). Greenpeace is calling for a worldwide ban on the release of any transgenic crop or seed and for governments to halt both commercial and experimental growing of genetically engineered crops.

On August 10, 2006, a federal court issued in *Center for Food Safety et al. v. Johanns et al.* the first ruling ever on biopharming, the controversial practice of genetically

TABLE 9.1

Bioengineered foods approved by the U.S. Food and Drug Administration, 2002–05

[By year and file number (BNF number)]

Food	Gene, gene product, or gene fragment	Source	Intended effect	Designation	FDA letter	FDA memo
Submissions completed in 2005						
BNF No. 98, submitted May 27, 2004 by Monsanto Company, for use in human food and animal feed Cotton	5-enolpyruvylshikimate-3-phosphate synthase (EPSPS)	Agrobacterium sp. strain CP4	Tolerance to the herbicide glyphosate	MON-88913-8	7-Mar-05	7-Mar-05
BNF No. 97, submitted March 30, 2004 by Monsanto Company, for use in human food and animal feed Corn	Cry3Bb1; 5-enolpyruvylshikimate-3-phosphate synthase (EPSPS)	Bacillus thuringiensis subsp. kumamotoensis; Agrobacterium sp. strain CP4	Resistance to corn rootworm; Tolerance to the herbicide glyphosate	MON 88017	12-Jan-05	5-Jan-05
BNF No. 94, submitted October 27, 2003 by Syngenta Seeds, Inc., for use in human food and animal feed Cotton	VIP3A protein	B. thuringiensis, strain AB88	Resistance to lepidopteran insects	Transformation event COT102	8-Jul-05	7-Jul-05
BNF No. 87, submitted August 10, 2004 by Monsanto Company, for use in human food and animal feed Corn	Dihydrodipicolinate synthase (cDHDPS)	Corynebacterium glutamicum	Increase lysine level for use in animal feed	REN-∅∅∅38-3 or maize event LY038	5-Oct-05	30-Sep-05
Submissions completed in 2004						
BNF No. 93, submitted June 30, 2003 by Mycogen Seeds c/o Dow AgroSciences LLC, for use in human food and animal feed Corn	Cry1F; phosphinothricin acetyltransferase (PAT)	Bacillus thuringiensis aizawai; Streptomyces hygroscopicus	Resistance to certain lepidopteran insects; Tolerance to the herbicide glufosinate-ammonium	Event TC6275	30-Jun-04	28-Jun-04
BNF No. 92, submitted March 18, 2003 by Mycogen Seeds c/o Dow AgroSciences LLC, for use in human food and animal feed Cotton	Cry1Ac; phosphinothricin acetyltransferase (PAT)	Bacillus thuringiensis subsp. kurstaki, Streptomyces viridochromogenes	Resistance to certain lepidopteran insects; Tolerance to the herbicide glufosinate-ammonium	Event 3006-210-23	3-Aug-04	28-Jul-04
BNF No. 90, submitted April 16, 2003 by Monsanto Company and KWS SAAT AG, for use in human food and animal feed Sugar beet	5-Enolpyruvylshikimate-3-phosphate synthase (EPSPS)	Agrobacterium sp. strain CP4	Tolerance to the herbicide glyphosate (N-phosphonomethyl glycine)	Event H7-1	17-Aug-04	7-Aug-04
BNF No. 85, submitted March 17, 2003 by Mycogen Seeds c/o Dow AgroSciences LLC, for use in human food and animal feed Cotton	Cry1F; phosphinothricin acetyltransferase (PAT)	Bacillus thuringiensis subsp. aizawai; Streptomyces viridochromogenes	Resistance to lepidopteran insects; Tolerance to the herbicide glufosinate-ammonium	Event 281-24-236	10-May-04	5-May-04

TABLE 9.1

Bioengineered foods approved by the U.S. Food and Drug Administration, 2002–05 [CONTINUED]

[By year and file number (BNF number)]

Food	Gene, gene product, or gene fragment	Source	Intended effect	Designation	FDA letter	FDA memo
BNF No. 84, submitted October 6, 2003 by Monsanto Company and Forage Genetics, for use in human food and animal feed						
Alfalfa	5-enolpyruvylshikimate-3-phosphate synthase (EPSPS)	*Agrobacterium* sp. strain CP4	Tolerance to the herbicide glyphosate	Event J101 and event J163	10-Dec-04	8-Dec-04
BNF No. 81, submitted December 11, 2003 by Mycogen Seeds c/o Dow AgroSciences LLC, for use in human food and animal feed						
Corn	Cry34Ab1, Cry35Ab1, phosphinothricin acetyltransferase (PAT)	*Bacillus thuringiensis* strain PS149B1; *Streptomyces viridochromogenes*	Resistance to coleopteran insects; Tolerance to the herbicide glufosinate-ammonium	DAS-59122-7	4-Oct-04	28-Sep-04
BNF No. 80, submitted June 28, 2002 by Monsanto Company, for use in human food and animal feed						
Wheat	5-enolpyruvylshikimate-3-phosphate synthase (EPSPS)	*Agrobacterium* sp. strain CP4	Tolerance to the herbicide glyphosate (N-phosphonomethyl-glycine)	MON 71800	22-Jul-04	22-Jul-04
Submissions completed in 2003						
BNF No. 86, submitted August 30, 2002 by Bayer CropScience USA LP, for use in human food and animal feed						
Cotton	phosphinothricin-N-acetyltransferase (PAT)	*Streptomyces hygroscopicus*	Tolerance to the herbicide glufosinate-ammonium	LLCotton25	2-Apr-03	5-Jun-03
BNF No. 79, submitted September 13, 2002 by Monsanto and The Scotts Company, for use in animal feed						
Creeping bentgrass	5-Enolpyruvylshikimate-3-phosphate synthase (EPSPS)	*Agrobacterium* sp. strain CP4	Tolerance to the herbicide glyphosate	Event ASR368	23-Sep-03	11-Sep-03
Submissions completed in 2002						
BNF No. 77, submitted April 30, 2001 by Monsanto Company, for use in human food and animal feed						
Oilseed rape (canola)	5-Enolpyruvylshikimate-3-phosphate synthase (EPSPS); Glyphosate oxidoreductase (GOX)	*Agrobacterium* sp. strain CP4, *ochrobactrum anthropi* strain LBAA	Tolerance to the herbicide glyphosate	GT200	5-Sep-02	4-Sep-02
BNF No. 74, submitted June 29, 2000 by Monsanto Company, for use in human food and animal feed						
Cotton	Cry2ab; Cry1ac	*Bacillus thuringiensis* subsp. *kumamotoensis*	Resistance to lepidopteran insects	15985	18-Jul-02	16-Jul-02

SOURCE: Adapted from "Completed Submissions Organized by Year and File Number (BNF No.)," in *List of Completed Consultations on Bioengineered Foods*, U.S. Food and Drug Administration, Center for Food Safety and Applied Nutrition, Office of Food Additive Safety, 2005, http://www.cfsan.fda.gov/~lrd/biocon.html#list (accessed November 7, 2006)

FIGURE 9.3

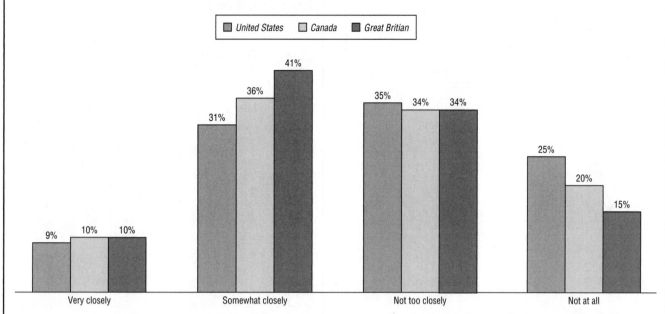

Public awareness among Americans, Canadians, and Britons of genetically modified food, 2005

AS YOU MAY KNOW, SOME FOOD PRODUCTS AND MEDICINES ARE BEING DEVELOPED USING NEW SCIENTIFIC TECHNIQUES. THE GENERAL AREA IS CALLED "BIOTECHNOLOGY" AND INCLUDES TOOLS SUCH AS GENETIC ENGINEERING AND GENETIC MODIFICATION OF FOOD. HOW CLOSELY HAVE YOU BEEN FOLLOWING THE NEWS ABOUT THIS ISSUE—VERY CLOSELY, SOMEWHAT CLOSELY, NOT TOO CLOSELY, OR NOT AT ALL?

SOURCE: Shelley Mika, "How Closely Have You Been Following the News about This Issue—Very Closely, Somewhat Closely, Not Too Closely, or Not at All?" in *Britons Show Distaste for Biotech Foods*, The Gallup Organization, October 2005, http://www.galluppoll.com:/content/Default .aspx?ci=19261&pg=1&t=rJL2jQQ5ucKNktt25CWMfTWqmbkp1zNUjsHGTtGZINlhHUWZw-N18rRBT.6ifGKEuur3ucbZRncqbsGAEv7XOO3SoUJ 8VT0c50f8dnl%2fvX4SNBGbNo9U9RmerLytvJxqixKDr46G3jcHMYoZjBsIpkZgDxvunU7q (accessed November 7, 2006). Copyright © 2006 by The Gallup Organization. Reproduced by permission of The Gallup Organization.

altering food crops to produce experimental drugs and industrial compounds. The court ruled that the U.S. Department of Agriculture violated the Endangered Species Act when it permitted the cultivation of drug-producing, GE crops in Hawaii. That same month the CFS reported in the press release "Unapproved, Genetically Engineered Rice Found in Food Supply" (August 18, 2006, http://www.centerforfoodsafety.org/GE_Rice_8.18 .06.cfm) that the USDA announced that an unapproved, GE rice was found contaminating commercial long-grain rice supplies. The presence of rice, which was genetically altered to survive application of the powerful herbicide glufosinate, in the food supply is illegal because it has not undergone USDA review for potential environmental impacts nor did the FDA review it for possible harm to human health.

On September 14, 2006, the CFS filed a petition with the USDA that sought to prevent after-the-fact approval of an illegal GE rice found in the world's food supply the previous month (http://www.centerforfoodsafety.org/ PR9_14_06.cfm). On October 12, 2006, the CFS and reproductive rights, animal welfare, and consumer protection organizations filed a petition with the FDA calling for a moratorium on the introduction of

food products from cloned animals (http://www.center forfoodsafety.org/cloning_petitionPR10.12.06.cfm). The petition also asked the FDA to establish mandatory rules for premarket food safety and environmental review of cloned foods.

U.S. Consumers' Opinions Vary

In 2005 the Gallup Organization found that only 9% of Americans were following the news about biotechnology "very closely." (See Figure 9.3.) A full 25% of Americans admitted that they were not following news about genetic engineering and genetic modification of food "at all," and an additional 35% said they did not follow such news "too closely." When queried about their support of or opposition to the use of biotechnology in agriculture and food production, equal proportions of respondents (45%) said they supported and opposed its use. (See Figure 9.4.)

In 2005 just one-third of Americans (33%) believed that foods produced using biotechnology pose a serious health hazard to consumers. (See Figure 9.5.) Fifty-four percent of respondents said they did not think GM foods pose a serious health hazard. There was less skepticism about GM foods and more support for them in the United States than in Canada or Great Britain.

FIGURE 9.4

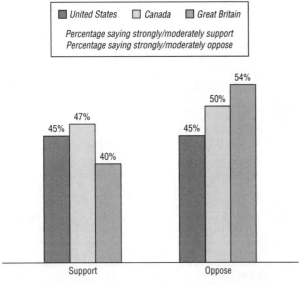

Public opinion among Americans, Canadians, and Britons on the use of biotechnology in agricultural food production, 2005

OVERALL, WOULD YOU SAY YOU STRONGLY SUPPORT, MODERATELY SUPPORT, MODERATELY OPPOSE, OR STRONGLY OPPOSE THE USE OF BIOTECHNOLOGY IN AGRICULTURE FOOD PRODUCTION?

■ United States ☐ Canada ■ Great Britain

Percentage saying strongly/moderately support
Percentage saying strongly/moderately oppose

SOURCE: Shelley Mika, "Overall, Would You Say You Strongly Support, Moderately Support, Moderately Oppose, or Strongly Oppose the Use of Biotechnology in Agriculture Food Production?" in *Britons Show Distaste for Biotech Foods*, The Gallup Organization, October 2005, http://www.galluppoll.com:/content/Default.aspx?ci=19261&pg= 1&t=rJL2jQQ5ucKNktt25CWMfTWqmbkp1zNUjsHGTtGZINlhHU WZw-N18rRBT.6ifGKEuur3ucbZRncqbsGAEv7XOO3SoUJ8VT0c50 f8dnl%2fvX4SNBGbNo9U9RmerLytvJxqixKDr46G3jcHMYoZjBsIpk ZgDxvunU7q (accessed November 7, 2006). Copyright © 2006 by The Gallup Organization. Reproduced by permission of The Gallup Organization.

FIGURE 9.5

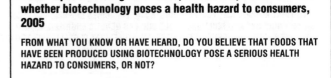

Public opinion among Americans, Canadians, and Britons on whether biotechnology poses a health hazard to consumers, 2005

FROM WHAT YOU KNOW OR HAVE HEARD, DO YOU BELIEVE THAT FOODS THAT HAVE BEEN PRODUCED USING BIOTECHNOLOGY POSE A SERIOUS HEALTH HAZARD TO CONSUMERS, OR NOT?

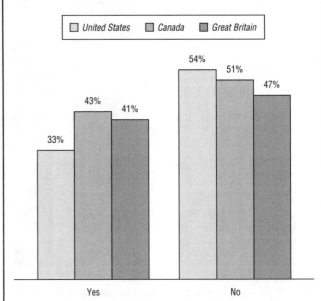

☐ United States ☐ Canada ■ Great Britain

SOURCE: Shelley Mika, "From What You Know or Have Heard, Do You Believe that Foods that Have Been Produced Using Biotechnology Pose a Serious Health Hazard to Consumers, or Not?" in *Britons Show Distaste for Biotech Foods*, The Gallup Organization, October 2005, http://www.galluppoll.com:/content/Default.aspx?ci=19261&pg=1&t =rJL2jQQ5ucKNktt25CWMfTWqmbkp1zNUjsHGTtGZINlhHU WZw-N18rRBT.6ifGKEuur3ucbZRncqbsGAEv7XOO3SoUJ8VT0c50f 8dnl%2fvX4SNBGbNo9U9RmerLytvJxqixKDr46G3jcHMYoZjBsIpk ZgDxvunU7q (accessed November 7, 2006). Copyright © 2006 by The Gallup Organization. Reproduced by permission of The Gallup Organization.

AN INTERNATIONAL FOOD FIGHT

Cartagena Protocol on Biosafety

In January 2000, after participating in the negotiation of the International BioSafety Protocol, a treaty developed by the United Nations Convention on Biological Diversity in which all signatories concurred that GM crops are significantly different from traditional crops, the United States refused to join other countries in signing. Known as the Cartagena Protocol on Biosafety, the treaty aims to ensure the safe transfer, handling, and use of living modified organisms—such as GE plants, animals, and microbes—across international borders. The protocol is also intended to prevent adverse effects on the conservation and sustainable use of biodiversity without unnecessarily disrupting world food trade. By participating in the treaty negotiations, the U.S. government formally acknowledged that GM crops are not, as many government agencies had previously maintained, "substantially equivalent" to traditional crops.

The Cartagena Protocol provides countries the opportunity to obtain information before new bioengineered organisms are imported. It acknowledges each country's right to regulate bioengineered organisms, subject to existing international obligations. It also aims to improve the capacity of developing countries to protect biodiversity. The protocol does not, however, cover processed food products or address the safety of GM food for consumption. Instead, the protocol is intended to protect the environment from the potential effects of introducing bioengineered products (referred to in the treaty as living modified organisms).

In September 2003 the Cartagena Protocol on Biosafety entered into force, empowering the more than one hundred countries that ratified the protocol to bar imports of live GMOs that they believe carry environmental or health risks. If the government of the importing country has concerns about safety, it can ask the exporting country to provide a risk assessment. The protocol also established a central online database of information on risks.

Although the protocol was acknowledged as an important first step in the protection of biodiversity, environmental advocacy groups including Greenpeace cautioned that there are still important unresolved issues for the international community, such as the amount of information required on shipments of GM crops. More important, Stephen Leahy reports in "Science: Bioterror Fears Dim Biotech Potential" (February 28, 2006, http://ipsnews.net/news.asp?idnews=32325) that the United States, Argentina, and Canada—the countries that produce about 90% of GE crops in the world—have not ratified the protocol.

Europe Bans Biotech Foods and Halts U.S. Trade

Although U.S. supermarket shelves remain stocked with GM foods, European consumers and farmers have nearly driven GE foods and crops out of the European Union (EU) market, the largest in the world. Europeans were so wary of GM foods that, beginning in 1998, the EU's European Commission and member states unofficially began a suspension on imports of biotech foods. At the onset of the ban, European Commission officials advised the United States they would resume the approval process once companies submitting applications agreed to abide by newly proposed revisions to the approval procedures before they become law. Although applicant companies agreed to comply voluntarily with the new procedures, the European Commission approval process did not resume as promised.

In June 1999 EU members called for an official moratorium on new approvals of agricultural biotech products. The EU Environmental Council contended that administration of new approvals should be linked to new labeling rules for biotech foods. Ministers from Denmark, France, Greece, Italy, and Luxembourg vowed to suspend approvals until new rules were established. A year later the EU environmental ministers moved to continue the moratorium at least until the European Commission prepared proposals for labeling and for tracing minute amounts of biotech products in foods such as vegetable and corn oils. The commission assured the United States that it would develop its proposal by the end of 2000 and promptly resume the approval process.

The traceability and labeling requirements were not presented until July 2001, when the European Commission promised to lift the moratorium within weeks. Frustrated by many delays and effects of these sanctions, the USDA determined that the EU moratorium on agricultural biotech products violated international law. The USDA argued in the fact sheet "Agricultural Biotechnology: WTO Case on Biotechnology" (September 2006, http://www.fas.usda.gov/itp/wto/eubiotech/factsheets/2006-09-28-biotech-wtocase.pdf) that:

- The EU "has pursued policies that undermine the development and use of agricultural biotechnology."

- In the late 1990s six member states (Austria, France, Germany, Greece, Italy, and Luxembourg) violated European law when they "banned imports of corn and rapeseed approved by the European Union," and the European Commission did not challenge the bans.

- In 1998 member states began to block EU regulatory approval for new agricultural biotech products. The moratorium prohibited most U.S. corn exports from entering Europe and violated EU law.

- The ban also breached WTO rules, which do not require automatic approval of biotech foods. The WTO requires that measures regulating imports be based on "scientific principles and evidence" and that countries operate regulatory approval procedures without "undue delay."

The United States asked the EU to apply a scientific, rules-based review and approval process to agricultural biotech product applications. The United States contended that it was not attempting to force European consumers to accept biotech foods or products. Instead, it denounced Europe's actions, which it considered whimsical rather than based on scientific, health, or environmental evidence, as depriving consumers of choice.

In 2004 Europe's moratorium on GM foods came to an end when the European Commission agreed to import worm-resistant GM corn known as Bt-11 developed by the Swiss firm Syngenta. In 2006 the World Trade Organization condemned the long approval process for GE products entering the European market as unscientific and determined that the process amounted to a trade embargo against agriculture producers in the United States, Canada, and Argentina. EU officials declined to appeal the ruling, but European consumers continued to view GE products with suspicion, and in early 2007 more than one million signatures had been gathered by Greenpeace in a petition drive calling for further investigation of biotech products entering Europe.

GE Foods in International Markets

Consumer rejection of GM foods has also been an issue to varying degrees in Japan, South Korea, Australia, New Zealand, India, and other nations. According to Roberto Verzola, in *The Genetic Engineering Debate* (July 2001, http://www.biotech-info.net/verzola_GE_debate.pdf), in May 2000 the Tokyo Grain Exchange soy futures market began to offer wholesale traders a choice of GE or conventional soybeans. On the first day of trading, 914,000 tons of conventional soybeans were purchased, compared with only 364,000 tons of GE soybeans. Beginning in 2001 Japan's Ministry of Health required all agricultural producers to screen imported

GM foods for potential food allergies and other health hazards and strengthened mandatory labeling rules on GM food ingredients. However, according to the Foreign Agricultural Service of the USDA, in spite of the testing and labeling guidelines U.S. agricultural products retained a strong market in Japan as of 2006. Annual exports to that country included 16 million metric tons of corn and 4.5 million metric tons of soybeans, much of it genetically engineered (http://www.fas.usda.gov/info/fasworldwide/2006/04-2006/JapanBiotech.pdf).

Consumer rejection of biotech products persists in Australia. In "Australia Struggles to Win Support for GMO Crops" (Reuters, March 9, 2005), Michael Byrnes notes that opposition to GM crops forced the country's poultry producers to stop feeding GM soy feed to the 450 million birds they sell each year. The Australian government has been actively trying to convince Australian consumers to welcome GM crops, and the country's farmers fear that they will suffer economic consequences if they cannot compete with farmers using GM organisms.

BIOTECHNOLOGY CORPORATIONS RESPOND TO CHALLENGES ABOUT GM FOODS

Corporations involved in genetic engineering research and production of GE crops and other products are alternately viewed as industry leaders providing valuable products and services or villains perpetrating economic, environmental, and health hazards on unsuspecting consumers throughout the world. Mounting consumer opposition to GM foods, especially outside the United States, prompted many of the giants in the agricultural biotech industry to scale back or even entirely halt production of some GM products.

For example, in 2000 Monsanto closed its NatureMark plant in Maine, a transgenic laboratory that produced Bt potatoes. Bt potatoes are gene-spliced with the soil bacterium *Bacillus thuringiensis* to repel the Colorado potato beetle. In November 1999 McCain's and Lamb-Weston, the two largest potato processors in North America, announced that they would no longer accept gene-altered potatoes; and the United States' leading potato purchasers, including McDonald's, Burger King, Frito-Lay, and Procter & Gamble, eliminated Bt potatoes from their french fries and potato chips. The Bt potatoes joined the growing list of GM foods abandoned by Monsanto. In 1996 Monsanto-Calgene Flavr Savr tomatoes were withdrawn from the market after they failed to meet expectations in terms of sales figures. Other major U.S. food corporations (including Gerber, Heinz, Mead-Johnson, and Frito-Lay) and several supermarket chains (such as Whole Foods, Wild Oats, and Genuardi's) announced plans to entirely forgo GM foods and products.

In May 2003 the World Agricultural Forum held the conference "A New Age in Agriculture: Working Together to Create the Future and Disable the Barriers" in St. Louis.

Presenters from the biotech firms Nestlé, Cargill, and Monsanto offered their intent to pursue free trade as well as their plans for using GM crops to help address world hunger.

To counter consumer opposition to GM foods and products, in 2000 the Council for Biotechnology Information aired television advertisements and launched a Web site to inform consumers of the potential benefits of GM foods. The media campaign emphasized that GM foods have been tested by U.S. government agencies and found to be safe; that biotechnology increases the nutritional content of foods, improves food quality, and can help feed the world's hungry; and that GM crops reduce the use, and environmental consequences of, toxic pesticides. In 2003 the council publicized in "Higher Corn Yields Are Making Ethanol More Energy Efficient" (http://www.whybiotech.com/index.asp?id=2213) the conclusions of the USDA study *The Energy Balance of Corn Ethanol: An Update* (July 2002, http://www.transportation.anl.gov/pdfs/AF/265.pdf). The study found that ethanol production is becoming increasingly energy efficient because corn yields are rising, less energy is required to grow it, and ethanol conversion technologies are becoming more efficient. These findings resulted from the use of GM corn crops. The council also offered as evidence the results of the National Center for Food and Agricultural Policy report *Plant Biotechnology: Current and Potential Impact for Improving Pest Management in U.S. Agriculture—An Analysis of 40 Case Studies* (June 2002, http://www.heartland.org/pdf/15786.pdf). The report found that "8 biotech cultivars adopted by U.S. growers increased crop yields by 4 billion pounds, saved growers $1.2 billion by lowering production costs and reduced pesticide use by 46 million pounds. These cultivars include insect resistant corn and cotton, herbicide tolerant canola, corn, cotton and soybean, and virus resistant papaya and squash. The adopted cultivars provided a net value of $1.5 billion."

In 2005 and 2006 the Council for Biotechnology Information continued to advance its premise that GM foods are safe by enlisting the support of other organizations including the American College of Nutrition, American Medical Association, International Society of Toxicology, and World Health Organization to attest to the safety of foods developed using biotechnology. The council asserts in "Food and Environmental Safety: Experts Say Biotech Food and Crops Are Safe" (2006, http://www.whybiotech.com/index.asp?id=2985) that more than 3,300 scientists, including twenty Nobel Prize winners, have signed a declaration in support of biotechnology and its safety. In "Protein-Rich Potato Could Help Combat Malnutrition in India" (2005, http://www.whybiotech.com/index.asp?id=4323), the council extols the virtues of genetically enhanced protein-rich potatoes, which could help to combat malnutrition in India.

PROMISE AND PROGRESS OF GENOMIC MEDICINE

Progress in understanding the genetic basis of disease has arrived at a rapid-fire pace. Genetic and genomic information gained from the Human Genome Project promises to revolutionize prevention and treatment of disease in the twenty-first century. Physicians will be able to accurately predict patients' risks of acquiring specific diseases and advise them of actions they may take to reduce their risks, prevent disease, and protect their health. There are equally promising therapeutic applications of genetic research including custom-tailored treatment that relies on knowledge of the patient's genetic profile and development of highly specific and effective medications to combat diseases.

According to Dave Carpenter in "Biotech: Practical Genomics" (*Hospitals and Health Networks*, May 2003), the genomics revolution has already arrived in some U.S. hospitals. Carpenter finds more than just the traditional genetic screening of newborns. He describes cardiovascular patients screened via sophisticated genetic analysis technology that employs computer software to determine patients' risk of complications, such as deep-vein thrombosis (blood clots in the leg veins).

The use of genetic analysis technology illustrates what industry observers call the shift from reactive patient care to predictive, preventive, and personalized care. Treatment can begin earlier and medication will be custom tailored for each patient. Physicians, administrators, and consumers expect that such genetic applications will prevent costly and potentially adverse surgical complications.

Molecular Farming Harvests New Drugs

Molecular farming or biopharming runs the gamut from tobacco plants harboring drugs to treat acquired immune deficiency syndrome (AIDS) to plants intended to yield fruit-based hepatitis vaccines. An example of a plant with the potential to produce a GM pharmaceutical is corn grown by the researcher Andy Hiatt at Epicyte Pharmaceutical in San Diego. According to Margot Roosevelt, in "Cures on the Cob" (*Time*, May 19, 2003), a human gene that codes for an antibody to genital herpes—a sexually transmitted disease that affects about sixty million Americans—is being grown in the corn plants. Epicyte is also developing plant-grown spermicides and antibodies to combat respiratory viruses, treat Alzheimer's disease, and counter Ebola, should the virus be used as a weapon in an act of bioterrorism.

Opponents call molecular farming "Pharmageddon," and environmentalists fear that the artificially combined genes will have unintended, untoward consequences for the environment. Consumer advocates, wary about the proliferation of GM foods, dread the possibility that plant-grown drugs and industrial chemicals will end up in food crops. To prevent such a scenario, the FDA issued new regulations to safeguard the food supply during the last quarter of 2003.

Preclinical Disease Detection

At the close of the twentieth century technological advances offered opportunities to identify diseases at stages before they were visible, in terms of biochemical or symptomatic expression. Called preclinical detection, this ability to predict and as a result intervene to prevent or avert serious disease involves an understanding of three levels of disease detection involving genomes, transcriptomes (the transcribed messenger RNA complement), and proteomes (the full range of translated proteins).

In "An Initial Map of Insertion and Deletion (INDEL) Variation in the Human Genome" (*Genome Research*, September 2006), Ryan E. Mills et al. describe a new kind of computer-based analysis to look for a type of genetic variation called an insertion and deletion (INDEL) polymorphism and the development of an initial map of human INDEL variation that contains 415,436 unique INDEL polymorphisms. In an INDEL variation, building blocks are added or deleted, not just switched on a one-for-one basis, and an insertion or deletion can involve thousands of blocks. Mills and his collaborators assert that INDELs represent about 25% of all genetic variations and believe that their work represents the dawn of a new era of predictive health.

Genes Explain Disease, Drug Resistance, and Treatment Failures and Successes

Research suggests that some people possess genes that enable the immune system to act unusually quickly. In "HLA and NK Cell Inhibitory Receptor Genes in Resolving Hepatitis C Virus Infection" (*Science*, August 6, 2004), Salim I. Khakoo et al. note this may explain how an estimated 20% of infected patients fend off or completely cure themselves of hepatitis C—a virus that causes serious and often fatal liver disease—without any medical treatment. The researchers posit that a specific gene combination allows the body to quickly let loose its frontline defense: natural killer cells. Natural killer cells are continually ready to counter an invading virus. Inhibitory receptors restrain natural killer cells between infections, to ensure they do not attack healthy tissue. Khakoo and his associates identified a particular gene combination that controls one inhibitory receptor, and the molecule attached to it was twice as common in recovered patients as in patients who remained infected with hepatitis C. To find the genes involved in this immune response, the researchers analyzed the DNA of 1,037 hepatitis C patients, 352 of whom spontaneously recovered. Khakoo et al. conclude that "[i]n the long term, whether we can use this information to modulate the body's immune system to improve therapeutics or vaccine design—that is the ultimate goal."

Amy Holleman et al., in "Gene-Expression Patterns in Drug-Resistant Acute Lymphoblastic Leukemia Cells and Response to Treatment" (*New England Journal of Medicine*, August 5, 2004), identify a set of genes linked to either resistance or sensitivity to the anticancer drugs commonly used to treat acute lymphoblastic leukemia. The researchers tested leukemia cells from 173 children newly diagnosed with leukemia for sensitivity to four chemotherapy drugs used in leukemia treatment. They found a particular group of genes that when present in leukemia cells determined their sensitivity or resistance to the drugs. The study also showed that these genes predicted treatment success or relapse in the 173 children as well as another group of 98 children with leukemia who were treated with the same drugs. Holleman and her coauthors assert that the presence or absence of these genes may explain why nearly 20% of children with leukemia do not respond to drug treatment.

In "EGFR Mutation and Resistance of Non-Small-Cell Lung Cancer to Gefitinib" (*New England Journal of Medicine*, February 24, 2005), Susumu Kobayashi et al. report that gene mutations explain why some lung cancer tumors become resistant to treatment with new cancer drugs meant to disrupt a molecular target that helps tumors grow. They find that mutations in the epidermal growth factor receptor gene are associated with favorable responses to treatment. They also find that the tumor stops responding to the cancer drugs if or when a secondary mutation in the same gene develops—three of six patients in the study who had this secondary mutation experienced a recurrence of their tumors. Kobayashi and his collaborators hypothesize that the anticancer drugs may give cancer cells that have the second mutation a growth advantage.

Marc Buyse et al., in "Validation and Clinical Utility of a 70-Gene Prognostic Signature for Women with Node-Negative Breast Cancer" (*Journal of the National Cancer Institute*, September 6, 2006), report that a new 70-gene assay will be able to more accurately predict the prognosis for women with breast cancer that has not yet spread to the lymph nodes. The technique accurately assesses whether the subjects—women fifty-five years or younger with node-negative breast cancer whose tumors were smaller than five centimeters in diameter—were low risk (a ten-year metastasis-free survival probability of greater than 90%) or high risk (a five-year probability of metastasis-free survival greater than 90%). This type of gene profiling will not only improve the accuracy of prognoses but also may help to predict which patients will benefit most from therapy.

Gene Therapy

Gene therapy aims to correct defective or faulty genes with one of several techniques. Most gene therapy involves the insertion of functioning genes into the genome to replace nonfunctioning genes. Other techniques entail swapping an abnormal gene for a normal one in a process known as homologous recombination, restoring an abnormal gene to normal function through selective reverse mutation, or changing the regulation of the gene—influencing the extent to which a gene is "turned on" or "turned off." Danny Penman Subtle notes in "Gene Therapy Tackles Blood Disorder" (October 11, 2002, http://www.newscientist.com/article.ns?id=dn2915) that gene repair of faulty messenger ribonucleic acid (mRNA) is being used by researchers at the University of North Carolina to treat blood disorders such as thalassemia and hemophilia as well as cystic fibrosis and some cancers. According to Bob Holmes, in "Gene Therapy May Switch Off Huntington's" (March 2003, http://www.newscientist.com/article.ns?id=dn3493), University of Iowa investigators report preliminary success with a related technique, known as RNA interference or gene silencing, to turn off production of an abnormal protein involved in Huntington's disease.

Early work in gene therapy focused on replacing a gene that was defective in a specific well-defined genetic disease such as cystic fibrosis. Recent research reveals that gene therapy technology may be more helpful in treating several nongenetic diseases for which there are no available effective treatments. Gene therapy clinical trials are currently under way for pancreatic cancer and sarcoma (a malignant tumor arising from nonepithelial connective tissues); end-stage (advanced) coronary artery disease, in which factors that improve blood supply to the heart may be lifesaving; and macular degeneration, for which loss of sight might be prevented.

To treat each of these disorders, the appropriate therapeutic genes are inserted by using in vivo (in a living organism, rather than the laboratory) gene therapy with adenoviral vectors. (Adenoviruses have double-stranded DNA genomes and cause respiratory, intestinal, and eye infections in humans.) A vector is a carrier molecule used to deliver the therapeutic gene to the target cells. The most frequently used vectors are viruses that have been genetically altered to contain normal human DNA. Researchers exploit viruses' natural abilities to encapsulate and deliver their genes to human cells. An unanticipated benefit of this type of gene therapy is that highly specific immunity could be induced by injecting into skeletal muscle DNA manipulated to carry a gene encoding a specific antigen. Because this form of therapy does not lead to integration of the donor gene into the host DNA, it eliminates potential problems resulting from disruption of the host's DNA, a phenomenon that was

FIGURE 9.6

Complex manufacture of a gene therapy product

Ex vivo transduced CD34+ cells expressing gammaC-R for X-SCID

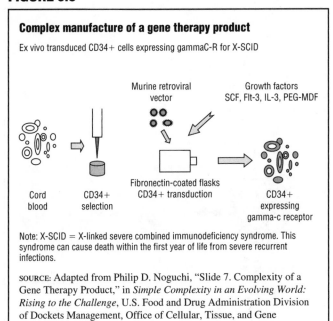

Note: X-SCID = X-linked severe combined immunodeficiency syndrome. This syndrome can cause death within the first year of life from severe recurrent infections.

SOURCE: Adapted from Philip D. Noguchi, "Slide 7. Complexity of a Gene Therapy Product," in *Simple Complexity in an Evolving World: Rising to the Challenge*, U.S. Food and Drug Administration Division of Dockets Management, Office of Cellular, Tissue, and Gene Therapies, October 2002, http://www.fda.gov/ohrms/dockets/ac/02/slides/3902s1-07-noguchi/sld007.htm (accessed November 8, 2006)

observed by investigators using ex vivo (outside a living organism, usually in the laboratory) gene therapy. Figure 9.6 shows how ex vivo cells are used to create a gene therapy product.

Figure 9.7 shows the steps involved in gene therapy using a retrovirus as the vector. In this process the vector discharges its DNA into the affected cells, which then begin to produce the missing or absent protein and are restored to their normal state. In this example the patient's own bone marrow cells are used as the vector to deliver severe combined immune deficiency–repaired genes to restore the function of the immune system.

Besides virus-mediated gene delivery, there are non-viral techniques for gene delivery. Direct introduction of therapeutic DNA into target cells requires large amounts of DNA and can be used only with certain tissues. Genes can also be delivered via an artificial lipid sphere, called a liposome, with a liquid core. This liposome, which contains the DNA, passes the DNA through the target cell's membrane. DNA can also enter target cells when it is chemically bound to a molecule that in turn binds to special cell receptors. Once united with these receptors, the therapeutic DNA is engulfed by the cell membrane and enters the target cell. Sylvia Pagán Westphal reports in "DNA Nanoballs Boost Gene Therapy" (May 12, 2002, http://www.newscientist.com/article.ns?id=dn2257) that researchers at Case Western Reserve University and Copernicus Therapeutics have created tiny liposomes able to transport therapeutic DNA through the pores of the nuclear membrane. Furthermore, Anil Ananthaswamy notes in "Undercover Genes Slip into the Brain" (March 20,

2003, http://www.newscientist.com/article.ns?id=dn3520) that researchers at the University of California at Los Angeles have successfully inserted genes into the brain using a liposome. This is a significant accomplishment because, previously, viral vectors were too large to cross the blood-brain barrier. The ability to transfer genes into the brain bodes well for patients suffering from neurological disorders such as Parkinson's disease.

In 1999 the death of the American teenager Jesse Gelsinger after he participated in a clinical gene therapy trial for ornithine transcarbamylase deficiency shocked and saddened the scientific community and diminished enthusiasm for technology that promises that healthy genes can replace faulty ones. Since then other adverse outcomes have tempered some initial success stories. In "Scientists Use Gene Therapy to Cure Immune Deficient Child" (*British Medical Journal*, July 6, 2002), Judy Siegel-Itzkovich notes that in 2002 an international team of scientists reported curing a child with severe combined immunodeficiency using gene therapy. The child spent the first seven months of her life inside a plastic bubble to protect her from all disease-causing agents because she totally lacked an immune system. After suppressing her defective bone marrow cells, the researchers introduced, using a GE virus, a healthy copy of the gene she was missing (for adenosine deaminase) into her purified bone marrow stem cells. The baby recovered quickly; within a few weeks she was no longer in isolation and went home healthy. Researchers credited the gene alteration treatment with the cure.

Two other children with similar conditions who received comparable gene therapy subsequently developed conditions resembling leukemia. As a result, in January 2003 the FDA temporarily stopped all gene therapy trials using retroviral vectors in blood stem cells. The FDA reconvened its Biological Response Modifiers Advisory Committee in February 2003 to determine whether to permit retroviral gene therapy trials for treatment of life-threatening diseases to proceed with additional safeguards. The committee indefinitely suspended such trials, and as of early 2007 they had not resumed in the United States.

In 2004 the Biological Response Modifiers Advisory Committee was renamed the Cellular, Tissue, and Gene Therapies Advisory Committee to describe more accurately the areas for which the committee is responsible. The FDA described the function of the renamed committee as evaluating "data relating to the safety, effectiveness, and appropriate use of human cells, human tissues, gene transfer therapies and xenotransplantation products" (September 24, 2004, http://www.fda.gov/cber/advisory/ctgt/ctgtchart.htm). Xenotransplantation is any procedure that involves transplantation, implantation, or infusion into a human recipient of live cells, tissues, or organs

FIGURE 9.7

Fundamentals of gene therapy

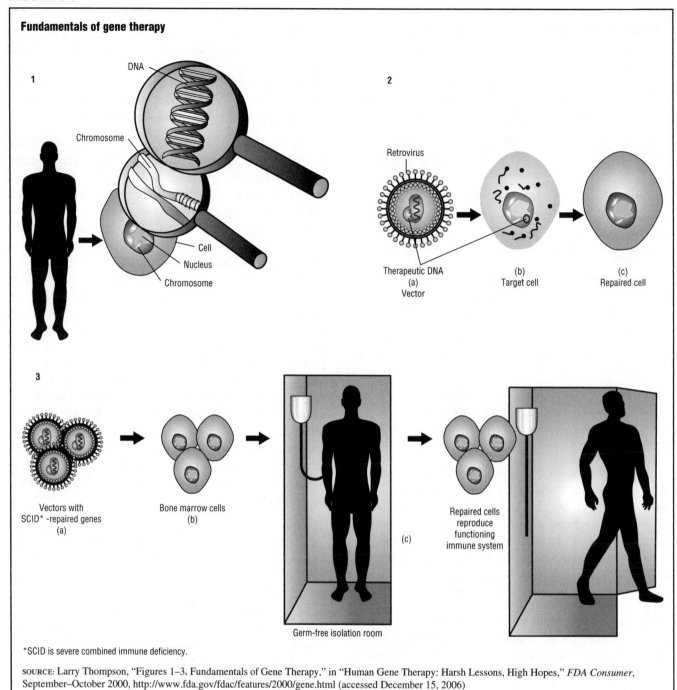

1

DNA

Chromosome

Cell

Nucleus

Chromosome

2

Retrovirus

Therapeutic DNA
(a)
Vector

(b)
Target cell

(c)
Repaired cell

3

Vectors with
SCID* -repaired genes
(a)

Bone marrow cells
(b)

(c)

Germ-free isolation room

Repaired cells
reproduce
functioning
immune system

*SCID is severe combined immune deficiency.

SOURCE: Larry Thompson, "Figures 1–3, Fundamentals of Gene Therapy," in "Human Gene Therapy: Harsh Lessons, High Hopes," *FDA Consumer*, September–October 2000, http://www.fda.gov/fdac/features/2000/gene.html (accessed December 15, 2006)

from a nonhuman animal source; or human body fluids, cells, tissues, or organs that have had any contact with live nonhuman animal cells, tissues, or organs. It has been used experimentally to treat certain diseases such as liver failure and diabetes, where there are insufficient human donor organs and tissues to meet demand.

The FDA and the National Institutes of Health (NIH) jointly oversee regulation of human gene therapy in the United States. The FDA focuses on ensuring that manufacturers produce quality, safe gene therapy products and that these products are adequately studied in human

subjects. The NIH evaluates the quality of the science involved in human gene therapy research and funds the laboratory scientists involved in development and refinement of gene transfer technology and clinical studies.

RECENT ADVANCES IN GENE THERAPY. Andrew Pollack reports in "Method to Turn Off Bad Genes Is Set for Tests on Human Eyes" (*New York Times*, September 14, 2004) that the FDA granted Acuity Pharmaceuticals permission in September 2004 to conduct the first human test of RNA interference (called RNAi) on patients suffering from macular degeneration—a deterioration of the retina that is the leading cause of blindness in older adults. The disease was chosen because the RNA can be directly injected into the eye, overcoming problems associated with delivering the RNA to the affected cells. Although RNAi has demonstrated efficacy in the laboratory, it is not yet known whether it will work in people. Other techniques once considered promising ways to turn off genes have not produced effective drugs. Other companies are investigating the use of RNAi to treat Huntington's and Parkinson's diseases, hepatitis C, and human immunodeficiency virus (HIV; the virus that causes acquired immunodeficiency syndrome). Pollack notes that Natasha J. Caplen, a gene therapy expert with the National Cancer Institute, said the FDA had not yet determined how it would regulate RNAi drugs. Until such regulations are in place, companies will be unable to begin clinical trials.

H. Bobby Gaspar et al., in "Gene Therapy of X-Linked Severe Combined Immunodeficiency by Use of a Pseudo-typed Gammaretroviral Vector" (*The Lancet*, December 18, 2004), report a successful gene therapy that corrects the cause of X-linked severe combine immunodeficiency (SCID-X1) and restores immunity. Bone-marrow stem cells were infused with human gamma-c cloned into an ape gamma-retroviral vector and then returned to the young patients, who ranged in age from four to thirty-three months. Gaspar and the other researchers conclude that "gene therapy for SCID-X1 is a highly effective strategy for restoration of functional cellular and humoral immunity."

According to Alice Dembner, in "Research to Unleash Gene Therapy on Arthritis" (*Boston Globe*, August 14, 2006), in 2006 scientists began testing gene therapies for osteoarthritis, a joint disease that afflicts more than twenty-one million Americans. One therapy involves injecting a gene into a diseased joint that will continuously pump medicine directly where it is needed. Another involves the use of GM cells that will pump proteins into the joint to stimulate growth of damaged cartilage.

In November 2006 researchers at the University of Pennsylvania's Abramson Family Cancer Research Institute reported promising results of a preliminary trial of a potential new gene therapy for HIV. In "Gene Transfer in Humans Using a Conditionally Replicating Lentiviral Vector" (*Proceedings of the National Academy of Sciences*, November 7, 2006), Bruce L. Levine et al. note that they removed immune cells from the patients and introduced a virus called a lentivirus into the cells. This change prevents HIV from reproducing and, in the laboratory, demonstrated the ability to fight HIV in cells that have not been treated. This first human study was conducted on patients with HIV infections that have been treatment resistant, and it showed that the treatment was safe and effective.

Genetically Enhanced Athletes

In "Gene Therapy May Be Up to Speed for Cheats at 2008 Olympics" (*Nature*, December 6, 2001), David Adams reports that gene therapy may enable athletes to genetically modify themselves to boost their performances. Athletes might target performance-enhancing genes such as those encoding growth factors capable of building muscle strength or widening blood vessels, or a hormone called erythropoietin that increases the number of oxygen-carrying red blood cells. Furthermore, the researchers Adams interviewed said such modifications might be impossible to detect and that artificial genes "can and most likely will be abused by healthy athletes as a means of doping." The International Olympic Committee has established an advisory group to monitor progress in gene therapy and to prevent a GM athlete from competing in the 2008 Olympic Games in Beijing.

MORE APPLICATIONS OF GENETIC RESEARCH

Most applications of genetic biotechnology are scientific, agricultural, and medical. However, geneticists are also engaged in product research and development of related technology and in legal determinations. Involvement with legal matters and the criminal justice system often takes the form of DNA profiling, also known as DNA fingerprinting. Because every organism has its own unique DNA, genetic testing can definitively determine whether individuals are related to one another and whether DNA evidence at a crime scene belongs to a suspect. It can also accurately identify a specific strain of a bacterium.

One example of researchers' use of genetic profiling to identify disease-causing bacteria was reported by Rex Dalton in "Genetic Sleuths Rush to Identify Anthrax Strains in Mail Attacks" (*Nature*, October 18, 2001). Dalton describes how the anthrax attacks in the eastern United States during the first weeks of October 2001 spurred researchers to work quickly to identify the strains of bacteria involved. The researchers hoped to help identify the origin of the anthrax spores and presumably trace them to their source. Paul Keim, a geneticist at Northern Arizona University in Flagstaff, led the research that used amplified fragment length polymorphism DNA analysis and another test called multilocus variable-number tandem repeat analysis (used with microorganisms to examine the *Bacillus anthraces*).

Dalton notes that it took Keim's research team about twelve hours to analyze a single sample. Even though several samples were ultimately determined to have been derived from the virulent Ames strain, it was not a simple task to trace the Ames strain to a single source because it has been passed around the world by researchers. It had been commonly used in laboratory research to develop vaccines and tests after its original isolate was removed from a dead animal in the 1950s near Ames, Iowa.

Forensics

The seventeenth edition of the *Merck Manual of Diagnosis and Therapy* (2003) describes forensic genetics as using molecular genetic techniques to identify an individual's genetic makeup. Forensic genetics relies on the measurement of many different genetic markers, each of which normally varies from individual to individual, and may be used to determine whether two people are genetically related. For example, because half of a person's genetic markers come from the father and half from the mother, analyses of these DNA markers enable laboratory technologists to establish that one person is the offspring of another. Analysis of DNA derived from blood samples can determine whether the supposed parents of a particular child are actually the biological parents. DNA markers may also be used to identify a specimen and establish its origin—that is, definitively determine the individual from whom it came. Biopsies, pathologic specimens, and blood and semen samples can all be used to measure DNA markers.

Forensic investigations often involve analyses of evidence left at a crime scene such as trace amounts of blood, a single hair, or skin cells. Using the polymerase chain reaction (PCR), DNA from a single cell can be amplified to provide a sample quantity that is large enough to determine the source of the DNA. In PCR the double strand of DNA is denatured into single strands, which are placed in a medium with the chemicals needed for DNA replication. The single strands both become double strands, yielding twice the amount of the initial DNA sample. The double strands are once again denatured to form single strands, and the process is repeated until there is a sufficient quantity of DNA for analysis. Analysis is usually performed using gel electrophoresis, in which DNA is loaded onto a gel and an electrical current is passed through the DNA. Investigators are then able to observe larger molecules migrating more slowly than smaller ones. Examining variable-number tandem repeats (VNTRs) is a procedure that identifies the length of tandem repeats in an individual's DNA. VNTR is an especially useful technique in forensic genetics, because when it is performed in careful lab conditions the probability of two individuals having the exact same VNTR results is less than one in a million.

DNA evidence is preferred by forensic specialists because, even though fingerprints can often be erased or eliminated and hair color and appearance may be altered, DNA is immutable. It can be used to identify individuals with extremely high probability and is more stable than other biological samples such as proteins or blood groups. Forensic genetics can be a powerful tool and

has been used successfully to eliminate suspects and clear their names. Though not as useful in proving guilt, it provides solid and often convincing evidence when many alleles match.

The first admission of DNA evidence in criminal court occurred in *Florida v. Tommy Lee Andrews* (1987), when the state of Florida used it as part of the prosecution case to convict a suspect of a series of sexual assaults. In 1989 the Federal Bureau of Investigation (FBI) began accepting work from state forensic laboratories, and in 1996 the National Research Council published *The Evaluation of Forensic DNA Evidence*, which cited the FBI statistic that approximately one-third of primary suspects in rape cases are excluded by using DNA evidence.

The International Society for Forensic Genetics promotes scientific knowledge in the field of genetic markers analyzed for forensic purposes. Many Americans first became aware of the use and importance of DNA evidence during the sensational and widely publicized 1994 trial of O. J. Simpson for the murder of Nicole Brown Simpson and Ron Goldman. Nicole Simpson's blood was found in O. J. Simpson's vehicle and house, but the defense discredited the DNA results by claiming the police had conducted a sloppy investigation that caused the samples to be contaminated and, alternatively, that the blood was planted in an attempt to frame Simpson.

Nanotechnology

A nanometer—the width of ten hydrogen atoms laid side by side—is among the smallest units of measure. It is one-billionth of a meter, one-millionth the size of a pinhead, or one-thousandth the length of a typical bacterium. Figure 9.8 shows the incredibly tiny scale of nanometers. According to an NIH definition (June 12, 2000, http://www.becon.nih.gov/nstc_def_nano.htm), nanotechnology

> *involves research and technology development at the atomic, molecular, or macromolecular levels in the dimension range of approximately 1–100 nanometers to provide fundamental understanding of phenomena and materials at the nanoscale and to create and use structures, devices, and systems that have novel properties and functions because of their small and/or intermediate size. The novel and differentiating properties and functions are developed at a critical length scale of matter typically under 100 nm. Nanotechnology research and development includes control at the nanoscale and integration of nanoscale structures into larger material components, systems, and architectures. Within these larger scale assemblies, the control and construction of their structures and components remains at the nanometer scale.*

Nanotechnology is sometimes called molecular manufacturing because it draws from many disciplines

FIGURE 9.8

| Visual representation of the size of a nanometer |

Less than a nanometer
Individual atoms are up to a few angstroms, or up to a few tenths of a nanometer, in diameter.

Nanometer
Ten shoulder-to-shoulder hydrogen atoms span 1 nanometer. DNA molecules are about 2.5 nanometers wide.

Thousands of nanometers
Biological cells, like these red blood cells, have diameters in the range of thousands of nanometers.

A million nanometers
The pinhead sized patch of this thumb (circled in black) is a million nanometers across.

Billions of nanometers
A two meter tall male is two billion nanometers tall.

SOURCE: Ivan Amato, "The Incredible Tininess of Nano," in *Nanotechnology: Shaping the World Atom by Atom*, National Science and Technology Council, Committee on Technology, Interagency Working Group on Nanoscience, Engineering, and Technology, 1999, http://www.wtec.org/loyola/nano/IWGN .Public.Brochure/IWGN.Nanotechnology.Brochure.pdf (accessed December 15, 2006)

(including physics, engineering, molecular biology, and chemistry) and considers the design and manufacture of extremely small electronic circuits and mechanical devices built at the molecular level of matter. K. Eric Drexler, the chief technical adviser of Nanorex, a company developing software for the design and simulation of molecular machine systems, coined the term *nanotechnology* in the 1980s to describe atomically precise molecular manufacturing systems and their products. He states in his book *Engines of Creation* (1986) that it is an emerging technology with the potential to fulfill many scientific, engineering, and medical objectives.

Researchers anticipate a myriad of practical applications of nanotechnology such as home food-growing machines that could produce virtually unlimited food supplies and chip-sized diagnostic devices that would revolutionize the detection and management of illness. Investigators envision computer-controlled molecular tools much smaller than a human cell and constructed with the accuracy and precision of drug molecules. Such tools would enable medicine to intervene in a sophisticated and controlled way at the cellular and molecular level—removing obstructions in the circulatory system, destroying cancer cells, or assuming the function of organelles such as the mitochondria. Other potential uses of nanotechnology in medicine include the early detection and treatment of disease via exquisitely precise sensors for use in the laboratory, clinic, and in the human body, plus new formulations and delivery systems for pharmaceutical drugs. In "Nanocontainers Deliver Drugs Directly to Cells" (*Scientific American*, April 28, 2003),

Sarah Graham describes how researchers at McGill University in Montreal have developed tiny drug delivery vehicles that are able to pass through the cell wall of a rat and even penetrate some cell parts such as the mitochondria and Golgi apparatus. Such highly refined and specific drug delivery systems may enable physicians to administer smaller doses of toxic medications more safely. In the not-too-distant future, nanorobots may act as programmable antibodies. As disease-causing bacteria and viruses mutate to elude medical treatments, the nanorobots could be reprogrammed to selectively seek out and destroy them, and others might be programmed to identify and eliminate cancer cells, leaving normal cells unharmed.

The technology may also be used to develop immediately compatible, rejection-resistant implants made of high-performance materials that respond as the body's needs change. Nanotechnology will lead to new biomedical therapies as well as prosthetic devices and medical implants. Some of these will help attract and assemble raw materials in bodily fluids to regenerate bone, skin, or other missing or damaged tissues. Nanotubes that act like tiny straws could conceivably circulate in a person's bloodstream and deliver medicines slowly over time or to highly specific locations in the body.

NEW DEVELOPMENTS IN NANOMEDICINE. Rick Weiss describes in "Nanomedicine's Promise Is Anything but Tiny" (*Washington Post*, January 30, 2005) some recent applications of nanotechnology to medicine, including:

• Quantum dot diagnostics—tiny bits of silicon just a few atoms in diameter—are being used to track the movement of substances in cells and to identify diseases in blood or other tissue.

- Tiny amounts of amino acids adhering to nanofibers that help nerve cells to heal and grow are delivered, suspended in a liquid gel called a nanogel.

- Photo-thermal nanoshells—gold-coated spheres just 130 nanometers in diameter absorb near infrared light and can be inserted deep into the body to the site of a tumor and heated to 122 degrees Fahrenheit, frying the tumor but not the surrounding tissue.

Another diagnostic application of nanomedicine is detailed by Dimitra G. Georganopoulou et al. in "Nanoparticle-Based Detection in Cerebral Spinal Fluid of a Soluble Pathogenic Biomarker for Alzheimer's Disease" (*Proceedings of the National Academy of Science*, February 15, 2005). The Northwestern University researchers report how nanoscience enables ultrasensitive detection of a biomarker for Alzheimer's disease, a neurodegenerative dementia that afflicts an estimated four million people in the United States. Until recently, Alzheimer's disease could only be conclusively diagnosed after death, when brain tissue could be examined for the plaques and neurofibrillary tangles that are the hallmarks of the disease.

Specific peptides—amyloid beta-derived diffusible ligands (ADDLs)—are believed to be the causative agent in memory loss associated with Alzheimer's disease. The researchers developed a bio-barcode assay, which is 100,000 times to 1 million times more sensitive than other available tests in the detection of ADDLs in the brain linked to Alzheimer's disease. Georganopoulou et al. hope that detecting ADDLs and other protein markers at significantly lower concentrations than conventional tests will lead to earlier diagnosis and intervention as well as development of new therapies for Alzheimer's and other diseases.

The editorial "Nanomedicine: A Matter of Rhetoric?" (*Nature Materials*, April 1, 2006) states that in 2006 an estimated 130 nanotech-based drugs and delivery systems were being developed worldwide. According to the *Nanomedicine Roadmap Initiative* (January 27, 2006, http://nihroadmap.nih.gov/nanomedicine/nanoinfomtg/pdf/NanoInfoMtg012706_Schloss.pdf), the NIH has established a national network of eight nanomedicine development centers, staffed by multidisciplinary biomedical scientific teams including biologists, physicians, mathematicians, engineers, and computer scientists. These centers aim to:

- Characterize quantitatively (through measurement) the physical and chemical properties of molecules and nanomachinery in cells

- Gain an understanding of the engineering principles used in living cells to build molecules, molecular complexes, organelles, cells, and tissues

- Use this knowledge of properties and design principles to develop new technologies, and engineer devices and hybrid structures for repairing tissues and for preventing and curing disease

CHAPTER 10
ETHICAL ISSUES AND PUBLIC OPINION

The institutions of genetic, scientific, and technical research, and the industries of genetic application, are relatively well organized and generously funded. Their imperatives are clear: push toward new knowledge and its applications. By contrast, our ethical, social discussion is unfocused, episodic, and scattered. We need to harness moral thinking to genetic technique. The need for organized, intelligent debate involving an active public and committed scientists has never been clearer.

—Everett Mendelsohn, "The Eugenic Temptation: When Ethics Lag behind Technology" (*Harvard Magazine*, March–April 2000)

Rapid advances in genetics and its applications pose new and complicated ethical, legal, regulatory, and policy issues for individuals and society. The issues society must consider include how to protect and manage genetic information and who should have access to it; the consequences of knowledge about personal genetic information for individuals; and the repercussions of genomic information for groups such as ethnic and racial minorities. For example, African-Americans were among the first to call for the inclusion of African-American genetic sequences in the human genome's template, and their concerns are vitally important because historically they have been victimized by "genetic inquiries" and research. Genetic discrimination poses a threat to members of minority groups, as well as to other workers, when employers are able to take adverse action based on applicants' or employees' asymptomatic genetic predisposition to, or probability of developing, a medical disorder.

Furthermore, many scientists, ethicists, and policy makers fear that without adequate protections genetic information may be used to deny individuals not only employment but also health care coverage or legal rights should genetic information be misused in the criminal justice system. They also worry that fear of genetic discrimination might deter participation in research or therapy. The American public shares these concerns about privacy and the management of genetic information. According to Christy White, John Meunier, and Gillian Steel Fisher, in "Public Perception of Genomics/Genetic Testing: CGAT Survey Results" (*Pharmaweek*, November 4, 2005), the 2005 Cogent Research Syndicated Genomics Attitudes and Trends Survey (CGAT), an annual survey that examines public awareness, understanding, favorability, and interest in genomics, found that more than two-thirds of the public believes insurance companies will try to use genetic information to deny health coverage, and 72% favor federal laws and regulations to protect genetic information. The CGAT survey also reports that most people strongly oppose government access to their genetic information.

In "Concerns in a Primary Care Population about Genetic Discrimination by Insurers" (*Genetics in Medicine*, May–June 2005), Mark Hall et al. also find considerable public concern about privacy and the potential for discrimination. Forty percent of those surveyed said the potential for discrimination affected their willingness to be tested, and agreed with the statement, "Genetic testing is not a good idea because you might have trouble getting or keeping your insurance."

In December 2004 the Genetics and Public Policy Center published *Reproductive Genetic Testing: What America Thinks* (http://www.dnapolicy.org/images/reportpdfs/ReproGenTestAmericaThinks.pdf), an analysis of survey data characterizing public awareness and knowledge, perceptions, and views about various aspects of genetic testing. The center is a part of the Phoebe R. Berman Bioethics Institute at Johns Hopkins University and is funded by the Pew Charitable Trusts. The center's mission is to create the environment and tools needed by decision makers in both the private and public sectors to carefully consider and respond to the challenges and opportunities that arise from scientific advances in genetics. Based on the largest public opinion survey ever conducted of American attitudes toward genetic testing, the research included twenty-one focus groups, sixty-two in-depth interviews, two surveys (one in 2002 and the other in 2004) with a combined sample size of more than 6,000 people, and both in-person and online town-hall meetings.

Researchers find that between 2002 and 2004 the proportion of Americans who believe spouses and extended family have the right to know genetic test results increased, whereas the proportion who believe employers and insurance carriers have a right to know decreased. About two-thirds of respondents in 2002 believed that the spouse of a person carrying a gene increasing the risk of disease had a right to know. In the 2004 survey more than 75% of survey respondents thought spouses had a right to know, and more than 50% indicated that extended family members also had a right to know. This contrasted sharply with support for disclosing genetic test results to insurance carriers and employers. In the 2004 survey only one-fifth of respondents believed insurance carriers had a right to such information, and fewer than 10% believed employers should be informed of test results.

A host of ethical issues related to health and medical care arise from increased availability of genetic information, such as the use of this information to guide reproductive decision making and the application of genetic engineering to reproductive technology. Reproductive applications are an especially emotionally charged topic as many antiabortion advocates staunchly oppose any action to alter the development or course of a naturally occurring pregnancy, such as in vitro fertilization and preimplantation intervention as well as the use of stem cells obtained from human embryos. Scientists, ethicists, and other observers also fear that genetic engineering technology will rapidly progress from enabling prenatal diagnosis and intervention to prevent serious disease to preconception selection of the traits and attributes of offspring—the creation of "designer babies."

These fears may be justified. Susannah Baruch and Kathy Hudson, in "The PGD Survey: Current Practice and Policy Implications" (*Fertility and Sterility*, September 2006), a survey of fertility clinics, find that nearly half the clinics that offer embryo screening permit couples to choose the gender of their child, and sex selection without any medical reason to warrant it was performed in 9% of all embryo screening performed in the United States in 2005.

Genetic testing challenges health care professionals to rapidly acquire new knowledge and skills to effectively use new technology and to assist their patients in making informed choices. The quality, reliability, and utility of genetic testing must be continuously reevaluated and regulated. Along with quality control measures, health care professionals and consumers must determine whether it is appropriate to perform tests for conditions for which no treatment exists and how to interpret and act on test results such as an individual's increased susceptibility to a disease associated with several genes and environmental triggers.

Philosophical, psychological, and spiritual considerations invite mental health professionals, ethicists, members of the clergy, and society overall to redefine concepts of human responsibility, free will, and genetic determinism. Behavioral genetics looks at the extent to which genes influence behavior and the human capacity to control behavior. It also attempts to describe the biological basis and heritability of traits ranging from intelligence and risk-taking behaviors to sexual orientation and alcoholism.

There are also environmental issues associated with the applications of genetic research, such as weighing the risks and benefits to human health and the environment of creating genetically modified animals, crops, and other products for human consumption. Legal and financial issues center on ownership of genes, deoxyribonucleic acid (DNA), and related data, property rights, patents, copyrights, and public access to research data and other genetic information.

HUMAN GENOME PROJECT CONSIDERS ETHICAL, LEGAL, AND SOCIAL ISSUES

Thoughtful review and analyses of policy issues and ethical considerations were deemed so pivotal to society and to the success of the Human Genome Project (HGP) that the Department of Energy and the National Institutes of Health dedicated 3% to 5% of the HGP's annual budget toward the study of the ethical, legal, and social issues (ELSI) arising from the availability of genetic information. Francis S. Collins et al. note in "New Goals for the U.S. Human Genome Project: 1998–2003" (*Science*, October 23, 1998) that the HGP's stated ELSI goals were to:

- Examine the issues surrounding the completion of the human DNA sequence and the study of human genetic variation

- Examine issues raised by the integration of genetic technologies and information into health care and public health activities

- Examine issues raised by the integration of knowledge about genomics and gene-environment interactions into nonclinical settings

- Explore ways in which new genetic knowledge may interact with a variety of philosophical, theological, and ethical perspectives

- Explore how socioeconomic factors and concepts of race and ethnicity influence the use, understanding, and interpretation of genetic information, the utilization of genetic services, and the development of policy

The HGP endeavor constitutes the world's largest bioethics program, and it has become a model for ELSI programs throughout the world.

Study Questions Impact of Disclosing Results of Genetic Testing

The National Human Genome Research Institute funds studies that consider the ethical and public policy implications of genetic testing and research. In one such study, "Genetic Testing for Alzheimer's Disease and Its Impact on Insurance Purchasing Behavior" (*Health Affairs*, March–April 2005), Cathleen D. Zick et al. look at how the results of genetic tests for adult-onset diseases might influence insurance purchases. They followed 148 healthy, normal people participating in a randomized clinical trial of genetic testing for Alzheimer's disease for one year after testing for the presence of a specific allele of the apolipoprotein E (apoE) gene, which if present increases the risk of developing Alzheimer's disease. Zick and her coauthors find that subjects who tested positive were 5.76 times more likely to have purchased or increased the scope of their long-term care insurance than those who tested negative for apoE. Presumably, this is because they anticipated increased need for insurance to help defray some of the costs of long-term medical and nonmedical care, such as assistance with activities of daily living.

According to the researchers, this finding concerns insurers, who fear that when genetic testing for Alzheimer's risk assessment becomes common, it will trigger adverse selection in long-term care insurance. Adverse selection occurs when individuals take financial advantage of risk classification systems. In this instance, insurance companies expect that a disproportionate number of people at risk of using the insurance will choose to purchase it. Insurance companies contend that they, too, should have access to the results of genetic testing because if they are unaware of occurrences of adverse selection they could experience economic losses.

Advocates of antigenetic discrimination legislation, however, maintain that if genetic test results are shared with insurers, many consumers could be denied coverage or charged excessively high premiums. They believe that distinctions made on the basis of genetic information are unfair because genetic makeup, unlike personal behaviors that modify health risks, cannot be changed. Furthermore, researchers speculate that the fear of discrimination may lead people to decline to participate in genetic research and testing.

Promise and Potential Perils of Genomics

In "Biology's New Forbidden Fruit" (*New York Times*, February 11, 2005), Oliver Morton outlines the scientific and commercial possibilities of synthetic biology—the new technology that enables scientists to write genes and genomes from scratch as opposed to trading naturally occurring genes from organism to organism. Morton presents examples of the great promise in designing genomes, such as the effort to produce the malaria drug artemisinin more cost effectively. He observes, however, that companies that synthesize genes on demand already exist, and he cautions that as this technology becomes increasingly affordable and available, the likelihood that someone may use it to create pathogens (infectious agents) increases. Furthermore, he acknowledges that diseases can be genetically engineered to be drug- and vaccine-resistant, warning that as a society we have not yet implemented a strategy to safeguard against the use of genetically engineered pathogens as weapons. Morton concludes that informed and concerned citizens can help to protect against misuse of genetic engineering technologies but adds that "to spur such debates in the wider public, biologists themselves will have to become more willing to think and talk about the ever more powerful technologies that they increasingly take for granted in the lab."

SURVEYS REVEAL SUPPORT FOR, AND CONCERN ABOUT, GENETIC RESEARCH AND ENGINEERING

Genetic research and engineering technologies promise to address some of the most pressing problems of the twenty-first century, such as cleaning the environment, feeding the world's hungry, and preventing serious diseases. However, like all new technologies, they pose risks as well as benefits. In general, Americans support advances in genetic research and technology, and they are optimistic that the outcomes will ultimately be used to reduce sickness and suffering, as opposed to generating legal and ethical controversies.

Most Americans are confident that genetic engineering holds great promise, and they support research and development of biotechnology, with one significant exception: cloning humans. Americans of all ethnicities, educational attainment, and political persuasions remain staunchly opposed to human cloning. Although there is some support for therapeutic, as opposed to reproductive, cloning initiatives, many Americans do not even endorse cloning domestic animals. Thirty-four percent of Americans surveyed by the Gallup Organization in 2003 said they thought cloning should be allowed for research purposes; however, 59% held that cloning should be banned. (See Table 10.1.) Americans have also rejected businesses engaged in commercial cloning activities. For example, the California-based Genetic Savings and

TABLE 10.1

FIGURE 10.1

Public opinion on a human cloning ban, 2003

AS YOU MAY KNOW, CONGRESS IS CONSIDERING SEVERAL PROPOSALS TO BAN HUMAN CLONING. WHICH OF THE FOLLOWING POSITIONS DO YOU MOST AGREE WITH—HUMAN CLONING SHOULD NOT BE BANNED, ONLY HUMAN CLONING THAT LEADS TO THE BIRTH OF A HUMAN SHOULD BE BANNED, BUT CLONING FOR PURPOSES OF LABORATORY RESEARCH SHOULD BE ALLOWED, OR ALL HUMAN CLONING SHOULD BE BANNED?

	Should not be banned	Should be allowed for purposes of research	Should be banned	Other (vol.)	No opinion
2003 Jan 13–16	4%	34	59	1	2

SOURCE: "As you may know, Congress is considering several proposals to ban human cloning. Which of the following positions do you most agree with—[ROTATED: human cloning should NOT be banned, only human cloning that leads to the birth of a human should be banned, but cloning for purposes of laboratory research should be allowed, or all human cloning SHOULD be banned]?" in *Cloning*, The Gallup Organization, January 2003, http://www.galluppoll.com:/content/Default.aspx?ci=6028&pg=1&t=tpkBD8 VGpZY%2fwOScYiI3tQAQNJ57o6ILWRxHFxGhFbRI.vkgYUEYW%2fl9q D-SKG-e7YUckhFo-B8sBSX6vc5YuQOCxedzjd0eZTR%2fKHLU5FM 9dobqyU4kl-itNIr7xDK-iyTsEostrJKXl%2fCjIrW2TyxXZD-TH4M%2f (accessed November 16, 2006). Copyright © 2006 by The Gallup Organization. Reproduced by permission of The Gallup Organization.

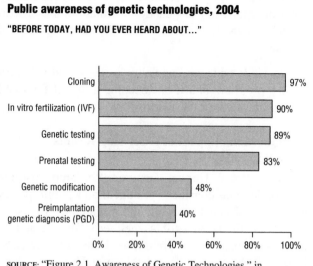

Public awareness of genetic technologies, 2004

"BEFORE TODAY, HAD YOU EVER HEARD ABOUT..."

Cloning	97%
In vitro fertilization (IVF)	90%
Genetic testing	89%
Prenatal testing	83%
Genetic modification	48%
Preimplantation genetic diagnosis (PGD)	40%

SOURCE: "Figure 2.1. Awareness of Genetic Technologies," in *Reproductive Genetics Testing: What America Thinks*, Genetics and Public Policy Center at the Johns Hopkins University, 2004, http://www .dnapolicy.org/images/reportpdfs/ReproGenTestAmerica Thinks.pdf (accessed November 16, 2006). Reproduced with permission of the Genetics and Public Policy Center, Washington, D.C.

Clone, Inc., which launched a pet cloning business in 2004, closed in late 2006.

Arguments for and against other genetic engineering applications, such as production of genetically modified food crops and the feasibility of stem cell research, arise from widely different philosophical and ethical perspectives on the new technology. Some of the fears about genetic diagnosis presume that the use of such technologies will be alien, impersonal, and technologically difficult. Scientists and advocates of applying genetic engineering to improve human life foresee a future in which parenting is guided by genetic testing, and new reproductive technologies are routine aspects of medical care in our lives. Bioethicists and others concerned about the implications of the routine use of this biotechnology contend that it is possible and advantageous to fund, discuss, and regulate genetics in the same way we currently consider environmental medicine, nutrition, and public health.

Americans Are Aware of Genetic Testing

According to the survey *Reproductive Genetics Testing*, researchers find relatively high levels of awareness about genetic technologies, with a majority of the American public reporting familiarity with cloning (97%), in vitro fertilization (90%), genetic testing (89%), and prenatal testing (83%). Nearly half of the survey respondents were acquainted with genetic modification, and 40% said they had heard of the relatively new technology of preimplantation genetic diagnosis (PGD). (PGD is genetic testing performed on embryos produced via in vitro

fertilization. It has been used by people at risk of transmitting a single gene disorder to their offspring as well as to detect chromosomal abnormalities in embryos of older women undergoing fertility treatment.) (See Figure 10.1.)

Knowledge and awareness of genetic testing vary by gender, educational attainment, and income, with women and those with higher education and income reporting the highest levels of knowledge about reproductive genetic technologies. (See Table 10.2.) Similarly, higher education is associated with accurate information about, and understanding of, the capabilities of genetic testing. More than two-thirds of respondents with postgraduate education were aware that genetic testing could be used to determine whether an individual is at increased risk of developing cancer compared to just half of those without a college education. Men were more likely than women (59.1% versus 54.9%), and whites more likely than African-Americans (57.9% versus 51.4%) to know the capabilities of genetic testing. (See Table 10.3.)

Approval of Genetic Testing Depends on How It Is Used

Americans' support for the use of reproductive genetic testing depends on how it is being used. Nearly three-quarters approve of the use of genetic testing to screen for fatal diseases and for tissue matching, and the majority supports testing to identify people at risk of developing diseases such as cancer. (See Table 10.4.) There is much less support for genetic testing for purposes of trait selection such as intelligence or strength. However, 51.3% of Americans believe that using prenatal genetic testing to select for gender is an appropriate use

TABLE 10.2

Public awareness of selected genetic technologies, by demographic characteristics, 2004

[Proportion of respondents who had heard about the following technologies prior to the interview]

Demographic characteristics	In vitro fertilization	Genetic testing	Prenatal genetic testing	Preimplantation genetic diagnosis	Genetic modification	Cloning
Total	**90.4**	**88.5**	**83.4**	**40.2**	**48.1**	**96.6**
Sex						
Men	88.4	86.0	78.0	36.3	51.6	96.3
Women	92.3	90.9	88.3	43.9	44.9	96.9
Age						
18–29	89.4	89.6	81.4	41.5	56.4	96.8
30–49	90.1	88.7	84.0	39.4	48.3	96.6
50+	91.2	87.8	83.8	40.4	43.1	96.6
Race/ethnicity						
White	92.8	91.0	85.1	41.4	50.4	97.7
Black	83.7	85.0	76.9	34.7	34.8	96.1
Hispanic	86.2	81.6	81.8	38.7	46.0	93.5
Religion						
Protestant[a]	91.3	90.3	83.3	41.0	42.6	98.3
Fund/Evangelical[b]	92.0	90.1	84.9	39.4	47.8	98.0
Catholic	92.7	88.9	84.8	39.5	50.8	96.8
Other Christian[c]	88.6	87.9	86.0	42.7	45.1	96.0
Other (non Christian)	88.3	82.3	76.5	30.7	55.6	93.3
No religion	87.0	87.0	81.2	43.0	56.1	93.9
Income						
Under 25k	87.0	83.2	80.1	40.5	42.7	93.9
25k–49k	90.0	89.4	82.3	38.4	46.3	97.4
50k–74.9k	92.7	92.0	84.4	39.8	52.4	98.7
75+k	95.8	94.2	91.4	44.2	58.9	98.4
Education						
No college	85.6	82.5	77.4	34.2	35.5	94.6
Some college	92.0	92.0	84.8	43.6	53.2	98.0
College	97.7	96.5	93.8	45.0	65.3	99.2
Post grad	97.8	96.2	92.2	53.1	68.8	98.4
Political affiliation						
Republicans	92.5	91.2	86.8	38.9	48.9	98.4
Other	83.9	80.6	75.9	36.1	39.3	92.7
Democrats	89.8	86.6	81.4	38.2	45.3	96.6

[a]Protestant includes respondents who self-identified as Protestant, excluding those who additionally self-identified as Fundamentalist or Evangelical.
[b]Fundamentalist/Evangelical includes all Protestant or other Christian respondents who additionally self-identified as Fundamentalist or Evangelical.
[c]Other Christian includes all who self-identified as other Christian, excluding those that additionally self-identified as Fundamentalist or Evangelical.

SOURCE: "Table 2.1. Proportion of Respondents Who Stated They Had Heard about the Following Technologies Prior to the Interview," in *Reproductive Genetics Testing: What America Thinks*, Genetics and Public Policy Center at the Johns Hopkins University, 2004, http://www.dnapolicy.org/images/reportpdfs/ReproGenTestAmericaThinks.pdf (accessed November 16, 2006)

of the technology. Support for different types of genetic testing varies slightly by gender, race, and ethnicity, but the greatest difference in support is by religion. Fundamentalist and Evangelical Christians were the least supportive of using reproductive genetic technologies for any purpose; Protestant respondents reported the highest levels of approval (30%) for genetic testing for traits than any other religious group. As educational attainment of respondents increased, so did support for prenatal genetic testing and testing to identify fatal diseases.

Overall, Americans are worried about how genetic technologies will be used. Three-quarters of those surveyed said they agreed with the statement, "Technology will inevitably lead to genetic enhancement and designer babies." (See Figure 10.2.) By contrast, other survey respondents, mostly male, contended that genetic technologies should be used

because intervention in reproduction will lead to tremendous improvements in human health. Table 10.5 shows that 55.6% of male respondents strongly agreed that reproductive genetic technology is potentially the next step in human evolution. Older adults, African-Americans, Democrats, and those without a religious affiliation were most likely to agree with this statement.

Eighty-one percent of respondents indicated considerable concern that the use of genetic technologies could result in discrimination against and stigmatization of people with disabilities. (See Figure 10.3.) The poll researchers note that women, people with less education, religious fundamentalists, and Evangelical Christians were most concerned about discrimination. Other concerns about the social consequences of genetic technologies include the fear that parents will be pressured to use technology or that

TABLE 10.3

Public awareness of the capabilities of genetic testing, 2004

[Percentage of correct responses to genetic testing questions]

Demographic characteristics	It is possible to test for certain kinds of cancer	It is not possible to test for intelligence or strength
Total	**56.9**	**22.1**
Sex		
Men	59.1	24.4
Women	54.9	20.0
Age		
18–29	56.5	28.3
30–49	57.8	26.7
50+	56.2	13.7
Race/ethnicity		
White	57.9	22.3
Black	51.4	16.0
Hispanic	55.4	23.1
Religion		
Protestant[a]	57.2	18.7
Fund/Evangelical[b]	55.2	19.0
Catholic	58.1	22.8
Other Christian[c]	53.9	20.5
Other (non Christian)	58.9	33.0
No religion	60.6	29.7
Income		
Under 25k	52.5	18.1
25k–49k	57.2	22.3
50k–74.9k	58.5	23.5
75+k	63.7	29.0
Education		
No college	49.5	16.9
Some college	62.2	23.1
College	63.3	28.5
Post grad	68.4	35.7
Political affiliation		
Republicans	57.4	23.3
Other	47.3	19.7
Democrats	56.0	19.5

[a]Protestant includes respondents who self-identified as Protestant, excluding those who additionally self-identified as Fundamentalist or Evangelical.
[b]Fundamentalist/Evangelical includes all Protestant or other Christian respondents who additionally self-identified as Fundamentalist or Evangelical.
[c]Other Christian includes all who self-identified as other Christian, excluding those that additionally self-identified as Fundamentalist or Evangelical.

SOURCE: "Table 2.2. Percentage of Correct Responses to Genetic Testing Questions," in *Reproductive Genetics Testing: What America Thinks* Genetics and Public Policy Center at the Johns Hopkins University, 2004, http://www.dnapolicy.org/images/reportpdfs/ReproGenTestAmericaThinks.pdf (accessed November 16, 2006)

there will be societal expectations that everyone should be flawless—free of imperfections, diseases, or disabilities. Seventy percent of those surveyed agreed that the ability to control human reproduction will lead to treating children like products. (See Figure 10.4.)

Americans were also concerned about legislation and oversight of reproductive genetic testing, balancing the wisdom of legislation and government oversight with the autonomy of individual and family decision making. The poll researchers find widespread concern about the potential for unregulated technology to get "out of control" (84%), but more than two-thirds of survey participants (70%) were

also skeptical about government regulators invading private reproductive decisions. About the same proportion (67%) said people should make their own decisions because the consequences of their choices were so intensely personal. More than three-quarters of respondents said the appropriate role of government was to oversee or track the effects of using genetic testing (76%); nearly three-quarters believed government should study the health effects of genetic testing; and 58% said the government should ensure equal access to this technology across demographic groups. (See Figure 10.5.)

ESTABLISHING PRIORITIES FOR GENETICS RESEARCH

In their essay "Some Gene Research Just Isn't Worth the Money" (*New York Times*, January 18, 2005), Keith Humphreys and Sally Satel describe the efforts of two geneticists, Kathleen Merikangas of the National Institute of Mental Health (NIMH) and Neil Risch of Stanford University, to develop a framework for establishing priorities for genetic research. The geneticists originally published their controversial stance in "Genomic Priorities and Public Health" (*Science*, June 4, 2004), in which they argued that genetic research should focus on diseases whose development and course are unaffected by personal behaviors or environmental influences, which may be easily modified. They cited autism (a complex developmental disability that typically appears during the first three years of life and affects three crucial areas of development: communication, social interaction, and creative or imaginative play), Type 1 diabetes, and Alzheimer's disease as examples of suitable candidates for genetic research.

Merikangas and Risch argued that disorders such as Type 2 diabetes, alcohol and nicotine addictions, and other disorders that may be averted by modifying personal behavior should not be the top priorities of genetic research because there are already effective interventions for these disorders. For example, maintaining a healthy weight, eating less, and exercising more can help prevent Type 2 diabetes, and environmental factors such as smoking bans, high sales taxes, and social pressures have helped reduce smoking and its related health problems. Addiction researchers, among others, hotly contested the geneticists' research agenda, arguing that even though some addictive behaviors might be modifiable through environmental approaches, the social and monetary costs associated with addiction are substantial enough to warrant genetic research dollars.

NOTED AUTHORS ADDRESS THE RISKS AND BENEFITS OF GENETIC ENGINEERING

In *Our Posthuman Future: Consequences of the Biotechnology Revolution* (2002), Francis Fukuyama, a professor of international political economy at Johns Hopkins

TABLE 10.4

Public opinion on prenatal genetic testing, by purpose of test and demographic characteristics, 2004

[Percentage of approval for prenatal genetic testing for selected purposes]

Demographic characteristics	Fatal	HLA match	Cancer	Sex	Traits
Total	**73.2**	**71.5**	**59.9**	**51.3**	**28.4**
Sex					
Men	73.7	68.8	64.1	56.9	33.8
Women	72.6	73.9	56.0	46.0	23.5
Age					
18–29	74.5	71.9	62.6	55.3	27.3
30–49	74.0	70.5	59.5	51.5	26.9
50+	71.5	72.3	58.7	48.6	30.7
Race/ethnicity					
White	73.8	71.5	58.9	49.8	25.2
Black	72.0	73.4	62.1	58.4	36.0
Hispanic	73.5	73.5	64.1	51.6	35.4
Religion					
Protestant[a]	76.9	76.3	62.2	53.1	30.1
Fundamentalist/Evangelical[b]	57.7	60.4	46.6	43.2	22.7
Catholic	74.9	75.9	60.2	50.0	27.1
Other Christian[c]	69.4	70.4	56.1	49.8	26.3
Other (non Christian)	79.4	64.3	66.1	54.8	31.1
No religion	82.9	74.4	71.7	58.8	34.4
Income					
Under 25k	71.1	72.0	59.1	53.2	34.6
25k–49k	73.4	72.0	61.7	48.5	26.4
50k–74.9k	72.3	71.2	59.3	53.2	25.6
75+k	77.9	69.5	58.1	50.7	22.9
Education					
No college	70.2	72.1	59.9	50.9	32.4
Some college	73.5	71.5	60.6	52.0	27.2
College	77.6	73.3	57.7	48.7	19.8
Post grad	79.9	65.3	61.1	55.5	26.5
Political affiliation					
Republicans	67.5	65.4	54.9	45.4	23.0
Other	69.1	67.0	57.3	48.8	28.9
Democrats	76.4	74.6	63.3	54.5	33.6
Aware prenatal testing	75.7	72.7	61.0	51.4	27.5

[a]Protestant includes respondents who self-identified as Protestant, excluding those who additionally self-identified as Fundamentalist or Evangelical.
[b]Fundamentalist/Evangelical includes all Protestant or other Christian respondents who additionally self-identified as Fundamentalist or Evangelical.
[c]Other Christian includes all who self-identified as other Christian, excluding those that additionally self-identified as Fundamentalist or Evangelical.
Note: HLA is human leukocyte antigen.

SOURCE: "Table 3.3. Percentage of Approval for Prenatal Genetic Testing by Purpose and Demographic Characteristics," in *Reproductive Genetics Testing: What America Thinks*, Genetics and Public Policy Center at the Johns Hopkins Universtiy, 2004, http://www.dnapolicy.org/images/reportpdfs/ReproGenTestAmericaThinks.pdf (accessed November 16, 2006)

University and a former member of the President's Council on Bioethics, cautions about the use and misuse of science and biotechnology. He observes that genetic engineering has gained popular acceptance as it is used to prevent or correct selected medical conditions, but he shares the widely held concerns about the unforeseeable results of gene manipulation and the potential for altering the complexion of society through the use of "enhancement technology" to customize the attributes of offspring. Fukuyama asserts his fear that those able to afford the genetic interventions to produce offspring who are smarter, stronger, more athletic, talented, and better looking will further widen the chasm between the economic classes. He writes that the ways in which society chooses to employ, regulate, and restrict genetic engineering may

challenge traditional concepts of human equality, changing existing understanding of human personality, identity, and the capacity for moral choice. Furthermore, he contends that genetic engineering technologies may afford societies new techniques for controlling the behavior of their citizens and could potentially overturn existing social hierarchies and affect the rate of intellectual, material, and political progress. Fukuyama predicts that genetic engineering and other applications of biotechnology have the potential to sharply alter the nature of global politics.

Despite his fears and cautions about the relatively recent ability to intentionally modify the human organism as opposed to waiting for evolution, Fukuyama does not consider it necessary or advisable to prohibit any actions that alter genetic codes. He would, however, start by

FIGURE 10.2

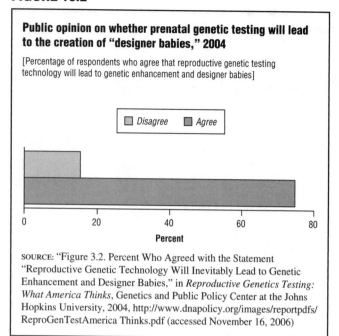

Public opinion on whether prenatal genetic testing will lead to the creation of "designer babies," 2004

[Percentage of respondents who agree that reproductive genetic testing technology will lead to genetic enhancement and designer babies]

SOURCE: "Figure 3.2. Percent Who Agreed with the Statement "Reproductive Genetic Technology Will Inevitably Lead to Genetic Enhancement and Designer Babies," in *Reproductive Genetics Testing: What America Thinks*, Genetics and Public Policy Center at the Johns Hopkins University, 2004, http://www.dnapolicy.org/images/reportpdfs/ ReproGenTestAmerica Thinks.pdf (accessed November 16, 2006)

TABLE 10.5

Public opinion on whether genetic technology is the next step in human evolution, by demographic characteristics, 2004

Demographic characteristic	Percent of those who agree or strongly agree
Total	53.6
Sex	
Men	55.6
Women	51.7
Age	
Age: 18–29	52.2
Age: 30–49	52.2
Age: 50+	55.9
Race/ethnicity	
White	52.4
Black	60.0
Hispanic	57.0
Religion	
Protestant[a]	56.7
Fundamentalist/Evangelical	41.0
Roman Catholic	53.8
Other Christian[c]	54.4
Other (non Christian)	56.7
No religion	60.9
Income	
Under 25k	57.1
25k–49k	52.5
50k–74.9k	50.7
75+k	52.0
Education	
No college	56.5
Some college	51.7
College	49.4
Post grad	51.6
Political affiliation	
Republicans	45.6
Other affiliation	52.2
Democrats	58.4

[a]Protestant includes respondents who self-identified as Protestant, excluding those who additionally self-identified as Fundamentalist or Evangelical.
[b]Fundamentalist/Evangelical includes all Protestant or other Christian respondents who additionally self-identified as Fundamentalist or Evangelical.
[c]Other Christian includes all who self-identified as other Christian, excluding those that additionally self-identified as Fundamentalist or Evangelical.

SOURCE: "Table 5.2. Reproductive Genetic Technology Is Potentially the Next Step in Human Evolution,"in *Reproductive Genetics Testing: What America Thinks*, Genetics and Public Policy Center at the Johns Hopkins University, 2004, http://www.dnapolicy.org/images/reportpdfs/ ReproGenTestAmericaThinks.pdf (accessed November 16, 2006)

banning reproductive cloning outright, to establish a precedent for political control over biotechnology. If society does not institute effective regulation, Fukuyama warns:

> We may be about to enter into a posthuman future, in which technology will give us the capacity gradually to alter [human] essence over time. . . . We do not have to regard ourselves as slaves to inevitable technological progress when that progress does not serve human ends. True freedom means the freedom of political communities to protect the values they hold most dear, and it is that freedom that we need to exercise with regard to the biotechnology revolution today.

Bill McKibben, the author of the renowned book *The End of Nature* (1989), which describes the consequences of damage done to the environment by overpopulation and global warming, also published *Enough: Staying Human in an Engineered Age* (2003), in which he warns that genetic engineering may not fulfill its promises. McKibben worries that society has been oversold on the potential benefits of genetic manipulation and that promises that it will make future generations healthier, smarter, happier, taller, thinner, better-looking, stronger, and saner may instead rob future generations of free will and freedom of choice. Like Fukuyama, McKibben is concerned that enabling the wealthy to custom-equip their offspring with good looks, high intelligence quotients, and athletic prowess will reinforce and deepen existing social class distinctions.

Although advocates claim it will prevent debilitating and fatal diseases and forestall death, McKibben believes that even the genetically enhanced will suffer. He fears that they will be beset by self-doubts, wondering if their achievements are their own or simply attributable to the geneticist. He wonders whether children whose parents have chosen to enhance them in the lab before birth will be able to make choices about their own lives, or will they just be making choices according to their prescripted genetic plans? McKibben contends that "the person left without any choice at all is the one you've engineered."

McKibben also argues against using germline engineering technologies to prevent diseases or germline gene therapy to treat diseases. Such therapies involve more

FIGURE 10.3

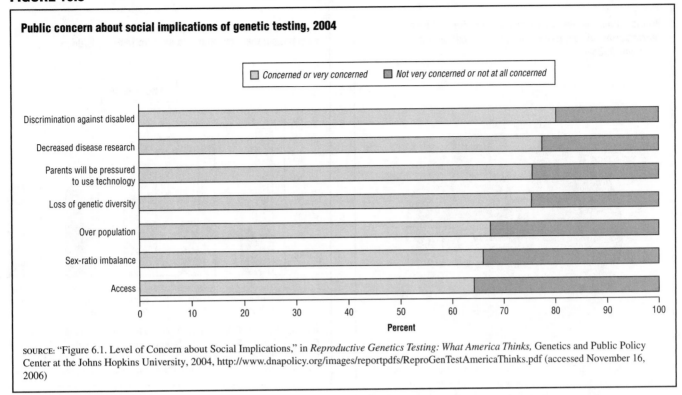

Public concern about social implications of genetic testing, 2004

☐ *Concerned or very concerned*   ☐ *Not very concerned or not at all concerned*

SOURCE: "Figure 6.1. Level of Concern about Social Implications," in *Reproductive Genetics Testing: What America Thinks*, Genetics and Public Policy Center at the Johns Hopkins University, 2004, http://www.dnapolicy.org/images/reportpdfs/ReproGenTestAmericaThinks.pdf (accessed November 16, 2006)

than simply altering the affected individuals' genes; they embed the genetic changes in their reproductive cells (sperm and eggs) so that the genetic alteration is heritable by all future generations. Furthermore, McKibben notes that society is as yet unable to distinguish between preventing disease and simply enhancing natural characteristics. Citing efforts to prevent dwarfism (short stature) and genetic enhancements to increase the height of a child who is naturally of short stature, he observes that "there's no obvious line between repair and improvement." He also expresses some skepticism about the feasibility of effectively regulating the distinction between repair and improvement in medical offices and clinics throughout the United States. McKibben asserts that the pace of change has accelerated such that decisions about how to use nanotechnology and other genetic engineering techniques must be made before their use is widely available and accepted. He fears that society may accept the view that humans are an "endlessly improvable species" and exhorts readers to conclude that human beings as currently constituted are good enough.

In *Playing God?: Human Genetic Engineering and the Rationalization of Public Bioethical Debate* (2002), the sociologist John H. Evans traces the public debate about human genetic engineering from the late 1950s through the mid-1990s. Evans believes that the debate has eroded over time from a substantive, rational discussion about the outcomes of human genetic engineering to a superficial dispute about the means to achieve a select

few results. He contends that by 1995 bioethicists no longer engaged in discussions of the larger, weightier, philosophical, and theological arguments about human genetic engineering. Instead, bioethics focused on resolving practical questions such as how medical decision making should occur and which parties should participate in the decision-making process. The widespread availability of genetic testing and counseling reinforced the argument that parents should be the ultimate arbiters of what is best for their offspring. Evans believes that early acceptance of these genetic engineering applications paved the way for societal acceptance of the inevitability of additional and potentially more controversial forms of human genetic engineering.

Evans concludes by calling for renewed debate about these significant ethical issues, recommending the establishment of separate groups to discuss the ends and means of human genetic engineering and cautioning that advisory commissions, to truly function representatively, should include lay people. He also urges the American public to become involved in human genetic engineering debates by voting and selecting legislators who respond to their constituencies when preparing and enacting resolutions and policies about the use of biotechnology.

The renowned bioethicist Arthur L. Caplan thinks that ethical implications of new knowledge in genetics are not at the forefront of the minds of professionals and health care consumers because of the uncertainty about

FIGURE 10.4

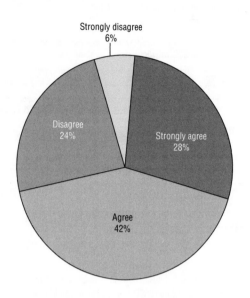

Public opinion on whether increased control of human reproduction will negatively affect the treatment of children, 2004

Strongly disagree
6%

Disagree
24%

Strongly agree
28%

Agree
42%

SOURCE: "Figure 6.2. The Ability to Control Human Reproduction Will Lead to Treating Children Like Products," in *Reproductive Genetics Testing: What America Thinks*, Genetics and Public Policy Center at the Johns Hopkins University, 2004, http://www.dnapolicy.org/images/reportpdfs/ReproGenTestAmericaThinks.pdf (accessed November 16, 2006)

FIGURE 10.5

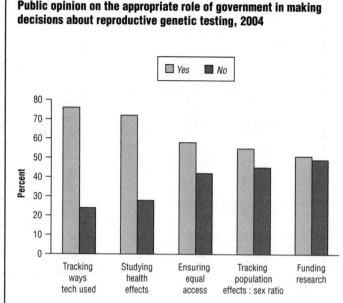

Public opinion on the appropriate role of government in making decisions about reproductive genetic testing, 2004

SOURCE: "Figure 8.4. Appropriate Role of Government?" in *Reproductive Genetics Testing: What America Thinks*, Genetics and Public Policy Center at the Johns Hopkins University, 2004, http://www.dnapolicy.org/images/reportpdfs/ReproGenTestAmericaThinks.pdf (accessed November 16, 2006)

how to use the new knowledge. In "If Gene Therapy Is the Cure, What Is the Disease?" (November 8, 2002, http://www.bioethics.net/articles.php?viewCat=6&articleId=58), Caplan, a strong supporter of the HGP, asserts that the greatest challenge to securing funding and support for genomic research is public fear of germline engineering—that is, manipulating the human genome to improve the human species—and he acknowledges that this fear is based on the historical reality of horrible eugenics (hereditary improvement of a race by genetic control) practices in Germany and other countries.

Caplan cites examples of genetic engineering in the United States such as the Repository for Germinal Choice in California, also known as the "Nobel Prize sperm bank," which solicits and stores sperm from men selected for their scientific, athletic, or entrepreneurial acumen. The banked sperm is available for use by women of high intelligence for the express purpose of creating genetically superior children. Caplan observes that there have been relatively few critics of this practice, whereas the mere suggestion of the possibility of directly modifying the genetic blueprint of gametes (sperm and eggs) has generated fiery debate in professional and lay communities. He contends that the history of eugenically driven social policy is reason enough to question and even protest the

actions of the Nobel Prize sperm bank but that it does not argue against allowing voluntary, therapeutic efforts using germline manipulations to prevent certain serious or fatal disorders from besetting future generations.

Caplan concludes that the decision to forgo germline engineering does not make ethical sense. He laments:

> [S]ome genetic diseases are so miserable and awful that at least some genetic interventions with the germline seem justifiable. . . . It is at best cruel to argue that some people must bear the burden of genetic disease in order to allow benefits to accrue to the group or species. At best, genetic diversity is an argument for creating a gamete bank to preserve diversity. It is hard to see why an unborn child has any obligation to preserve the genetic diversity of the species at the price of grave harm or certain death.

Caplan fears that choosing to refrain from efforts to modify the germline will result in lives sacrificed—that is, important benefits will be delayed or lost for people with disorders that might be effectively treated with germline engineering. Caplan recommends responding to justifiable concerns about the dangers and potential for abuse of new knowledge generated by the genome with frank, objective assessments of the appropriate goals of this application of biotechnology.

Caplan continues to exhort discussion of safeguards to prevent abuses of new knowledge and technologies. In a September 2, 2006, presentation, *Biobanking, Genomics, & Genetic Engineering: Where Are We Headed, and What*

Rules Should Take Us There? In addition to re-examining the question, "Should society limit how far we push genetic manipulation?" Caplan discussed concerns about biobanking—physical stores of human tissues and DNA, which often include the donor's personal information. He opined that biobanks offer tremendous potential benefits for selected patients, including those receiving organ transplants, but questioned whether there are controls in place to effectively prevent black market sales of human organs and other misuse of biobank tissues and confidential donor information.

IMPORTANT NAMES AND ADDRESSES

Alzheimer's Association
225 N. Michigan Ave., Seventeenth Floor
Chicago, IL 60601-7633
(312) 335-8700
1-800-272-3900
E-mail: info@alz.org
URL: http://www.alz.org/

American Board of Genetic Counseling
9650 Rockville Pike
Bethesda, MD 20814-3998
(301) 634-7315
FAX: (301) 634-7320
E-mail: info@abgc.net
URL: http://www.abgc.net/

American Cancer Society
1599 Clifton Rd. NE
Atlanta, GA 30329-4251
(404) 320-3333
1-800-ACS-2345
URL: http://www.cancer.org/

American Diabetes Association
1701 N. Beauregard St.
Alexandria, VA 22311
1-800-342-2383
E-mail: askada@diabetes.org
URL: http://www.diabetes.org/

American Heart Association
7272 Greenville Ave.
Dallas, TX 75231
1-800-242-8721
URL: http://www.americanheart.org/

American Parkinson Disease Association
135 Parkinson Ave.
Staten Island, NY 10305
(718) 981-8001
1-800-223-2732
FAX: (718) 981-4399
E-mail: apda@apdaparkinson.org
URL: http://www.apdaparkinson.org/

American Society of Gene Therapy
555 E. Wells St., Ste. 1100
Milwaukee, WI 53202
(414) 278-1341
FAX: (414) 276-3349
E-mail: info@asgt.org
URL: http://www.asgt.org/

American Society of Human Genetics
9650 Rockville Pike
Bethesda, MD 20814
(301) 634-7300
1-866-HUM-GENE
E-mail: society@ashg.org
URL: http://www.faseb.org/genetics/ashg/ashgmenu.htm

Center for Food Safety
660 Pennsylvania Ave. SE, Ste. 302
Washington, DC 20003
(202) 547-9359
FAX: (202) 547-9429
E-mail: office@centerforfoodsafety.org
URL: http://www.foodsafetynow.org/

Cold Spring Harbor Laboratory
One Bungtown Rd.
Cold Spring Harbor, NY 11724
(516) 367-8800
E-mail: info@cshl.edu
URL: http://www.cshl.org/

Council for Biotechnology Information
1225 Eye St. NW, Ste. 400
Washington, DC 20005
(202) 962-9200
URL: http://www.whybiotech.com/

Cystic Fibrosis Foundation
6931 Arlington Rd.
Bethesda, MD 20814
(301) 951-4422
1-800-344-4823
FAX: (301) 951-6378
E-mail: info@cff.org
URL: http://www.cff.org/

Genetics Society of America
9650 Rockville Pike
Bethesda, MD 20814-3998
(301) 634-7300
1-866-486-GENE
E-mail: society@genetics-gsa.org
URL: http://www.genetics-gsa.org/

Genome Programs of the U.S. Department of Energy Office of Science
1000 Independence Ave. SW
Washington, DC 20585
(202) 586-5430
URL: http://genomics.energy.gov/

Huntington's Disease Society of America
505 Eighth Ave., Ste. 902
New York, NY 10018
(212) 242-1968
1-800-345-4372
FAX: (212) 239-3430
E-mail: hdsainfo@hdsa.org
URL: http://www.hdsa.org/

The Institute for Genomic Research
9712 Medical Center Dr.
Rockville, MD 20850
(301) 795-7000
FAX: (301) 838-0208
URL: http://www.tigr.org/

International Genetic Epidemiology Society
Division of Biostatistics,
Washington University School
of Medicine
660 South Euclid Ave., Box 8067
St. Louis, MO 63110
(314) 362-3606
FAX: (314) 362-2693
URL: http://www.biostat.wustl.edu/genetics/iges/index.html

International Society for Forensic Genetics
c/o Institut für
Rechtsmedizin,Röntgenstrasse 23
D-48149 Münster, 10605 Germany
FAX: (+49) 251 83 55635

E-mail: niels.morling@forensic.ku.dk
URL: http://www.isfg.org/

**Muscular Dystrophy
Association–USA**
National Headquarters
3300 E. Sunrise Dr.
Tucson, AZ 85718-3208
1-800-572-1717
E-mail: mda@mdausa.org
URL: http://www.mdausa.org/

**National Down Syndrome
Society**
666 Broadway
New York, NY 10012
1-800-221-4602
FAX: (212) 979-2873
E-mail: info@ndss.org
URL: http://www.ndss.org/

National Human Genome Research Institute
National Institutes of Health
Bldg. 31, Rm. 4B09
31 Center Dr., MSC 2152
9000 Rockville Pike
Bethesda, MD 20892-2152
(301) 402-0911
FAX: (301) 402-2218
URL: http://www.genome.gov/

National Multiple Sclerosis Society
733 Third Ave.
New York, NY 10017
(212) 986-3240
1-800-344-4867
URL: http://www.nmss.org/

**National Tay-Sachs and Allied Diseases
Association**
2001 Beacon St., Ste. 204
Boston, MA 02135

1-800-906-8723
FAX: (617) 277-0134
E-mail: info@ntsad.org
URL: http://www.ntsad.org/

President's Council on Bioethics
1801 Pennsylvania Ave. NW, Ste. 700
Washington, DC 20006
(202) 296-4669
E-mail: info@bioethics.gov
URL: http://www.bioethics.gov/

**Wellcome Trust Medical Photographic
Library**
210 Euston Rd.
London, NW1 2BE United Kingdom
(+44) (0)20-7611-8348
FAX: (+44) (0)20-7611-8577
E-mail: medphoto@wellcome.ac.uk
URL: http://medphoto.wellcome.ac.uk/

RESOURCES

There are many published accounts of the history of genetics, but some of the most exciting versions were written by the pioneering researchers themselves. Although many sources were used to construct the historical overview and highlights contained in this book, James D. Watson's *The Double Helix: A Personal Account of the Discovery of the Structure of DNA* (1968), Francis H. C. Crick's memoir *What Mad Pursuit: A Personal View of Scientific Discovery* (1988), and Alfred H. Sturtevant's *A History of Genetics* (1965) provided especially useful insights. Thomas Hunt Morgan's *The Mechanism of Mendelian Heredity* (1915) and *The Theory of the Gene* (1926) offer detailed descriptions of groundbreaking genetic research. Ricki Lewis and Bernard Possidente offer more recent history in *A Short History of Genetics and Genetic Engineering* (2003, http://www.dna50.com/dna50.swf).

Charles Darwin's *On the Origin of Species by Means of Natural Selection* (1859) and *The Descent of Man, and Selection in Relation to Sex* (1871) provide historical perspectives on evolution. The discussion of nature versus nurture is drawn from Neil Whitehead and Briar Whitehead's *My Genes Made Me Do It* (1999), as well as Richard J. Herrnstein and Charles Murray's *The Bell Curve: Intelligence and Class Structure in American Life* (1994) and David Cohen's *Stranger in the Nest: Do Parents Really Shape Their Child's Personality, Intelligence, or Character?* (1999). Ethical issues arising from genetic research and engineering are analyzed in *Our Posthuman Future: Consequences of the Biotechnology Revolution* (2002) by Francis Fukuyama, *Playing God?: Human Genetic Engineering and the Rationalization of Public Bioethical Debate* (2002) by John H. Evans, and *Enough: Staying Human in an Engineered Age* (2003) by Bill McKibben.

The journals *Nature* and *Science* have reported every significant finding and development in genetics, and articles dating from 1953 from both publications are cited in this text, as are articles from *Scientific American*, *Nature Biotechnology*, *NewScientist.com*, and *e-biomed: The Journal of Regenerative Medicine*. Research describing genetic testing, disorders, and genetic predisposition to disease is reported in professional medical journals. Studies cited in this book were published in the *Archives of Disease in Childhood*, *Archives of Internal Medicine*, *British Medical Journal*, *Genetics in Medicine*, *Hospitals and Health Networks*, *Journal of Allergy and Clinical Immunology*, *Journal of the American Medical Association*, *New England Journal of Medicine*, and *Seminars in Respiratory and Critical Care Medicine*.

Ethical and psychological issues and the contributions of genetics to personality and behavior are examined in articles published in the *American Journal of Bioethics*, *Archives of General Psychiatry*, *British Journal of Psychiatry*, *European Psychologist*, *Health Affairs*, *Journal of the American Academy of Child and Adolescent Psychiatry*, *Journal of Consulting and Clinical Psychology*, *Journal of Educational Psychology*, *Journal of Personality and Social Psychology*, and *Psychological Review*.

Besides the U.S. daily newspapers and electronic media, accounts of genetics research and milestones in the Human Genome Project were reported in the magazines *Newsweek*, *New Yorker*, and *Time*.

The Human Genome Project Information Web site, which is operated by the U.S. Department of Energy, describes the ambitious goals and accomplishments of the Human Genome Project since its inception in 1990. The Environmental Genome Project was launched by the National Institute of Environmental Health Sciences. The National Institutes of Health provides definitions, epidemiological data, and research findings about a comprehensive range of genetic tests and genetic disorders.

Public opinion data from the following organizations was also very helpful: *Cogent Research Syndicated*

Genomics Attitudes and Trends Survey and the Gallup Organization. The Genetics and Public Policy Center is a part of the Phoebe R. Berman Bioethics Institute at the Johns Hopkins University. Additionally, many colleges, universities, medical centers, professional associations, and foundations dedicated to research, education, and advocacy about genetic disorders and diseases provided up-to-date information included in this edition.

INDEX

Cancer
  biotechnology, 134
  breast, 59(f5.10), 59(f5.11), 59(f5.12)
  cell cycle, 58f
  environmental exposure, 50
  genetics, 57–60
  treatment and genetic testing, 82
  twin studies, 46
Cancer Genome Project, 94
Canonical B-DNA double helix model, 11f
Caplan, Arthur L., 149–151
Caplen, Natasha J., 137
Carpenter, Dave, 133
Carrier identification, 72, 74(f6.2)
Cartagena Protocol on Biosafety, 130–131
Celera Genomics, 92, 93–94
Cell theory of heredity, 2
Cells
  cancer, 58f
  growth and reproduction, 21–24
  microscopy, 2
  structure and function, 20–21
  typical animal cell, 3f
Center for Food Safety (CFS), 125, 126, 129
Center for Food Safety et al. v. Johanns et al., 126, 129
Center for Inherited Disease Research, 92
Centromeres, 22, 23(f2.13)
CFS (Center for Food Safety), 125, 126, 129
Chakrabarty, Diamond v., 12
Chargaff, Erwin, 9
Chargaff's rules, 10f
Chase, Martha Cowles, 9
Check, Erika, 112
Chemical compounds, 103
Chemical Genomics Center, 103
Chemotherapy, 82
Children, 83
Chorionic villus sampling, 73–75
Chromosomes
  artificial, 12
  cytogenetics, 89, 90f
  disorders, 52
  gender determination, 27–28, 30(f2.23)
  gene crossovers, 24f
  heredity theory, 7–8
  human number estimates, 8
  meiosis, 26f
  mutations, 41
  structure, 22f
Chung, Young, 112
Cibelli, Jose B., 110
Classical genetics, 8–9
Clinical Laboratory Improvement Amendments (CLIA), 75
Clonaid, 109
Clone Registry database, 105
Cloning
  ethics, 113–114

first experiments, 12
food from cloned animals, 129
gene cloning, 105–107
human reproductive cloning, 109
legislation, 114–117
names, 106(f8.2)
public opinion, 117, 118(f8.8), 118(f8.9), 119f, 143–144, 144t
public policy, 112–113
reproductive, 107–109, 107f
therapeutic, 109–111
Codons, 19, 20(f2.8)
Cohen, David, 50
Collins, Francis S.
  ethics, 143–144
  evolution, 39
  genetic testing discrimination, 85
  Human Genome Project completion, 96, 98
  National Center for Human Genome Research, 91
Colorado Adoption Project, 48, 49–50
Commercialization of gene sequences, 91
Conservatism, 49–50
Consumer issues, 129, 130–132
"The Contribution of Genetics to Psychology Today and in the Decade to Follow" (Strelau), 49
Controversies
  genetically modified foods, safety of, 126–127
  terminator technology, 124–125
Corn, 124, 131, 132, 133
Corporations, biotechnology, 132
Correns, Karl Erich, 5
Council for Biotechnology Information, 132
Court cases
  Center for Food Safety et al. v. Johanns et al, 126, 129
  Diamond v. Chakrabarty, 12
  Florida v. Tommy Lee Andrews, 138
  Monsanto v. U.S. Farmers, 125
Creationism, 35, 36–39
Creighton, Harriet, 8
Crick, H. C., 9–10, 87
Crossovers, 22, 24f
Cutting DNA, 77, 81f
CVS (chorionic villus sampling), 73–75
Cystic fibrosis
  CFTR gene location, 60f
  chromosome 7 and the CFTR gene, 52(f5.1)
  gene therapy, 134
  genetic defects that cause, 61f
  genetics, 60–62
  Tsui, Lap-Chee, 89
Cytogenics, 89, 90f
Cytology, 2, 20

Dalton, Rex, 124, 137
D'Amour, Keven A., 110
Darrow, Clarence S., 35
Darwin, Charles, 2, 4, 33–34
Databases
  chemical genomics, 103
  Clone Registry database, 105
  genome sequencing, 93
  structure genomics, 103
Deaths, leading causes of, 54t
Definitions
  genome, 87
  nanotechnology, 138
Dembner, Alice, 137
Deoxyribonucleic acid. *See* DNA
Despande, Deepa, 110
Developing countries, 123, 124–125
Dexter, Michael, 94
Diabetes, 62–63, 63f, 63t
Diagnostic testing
  Alzheimer's disease, 56–57
  nanotechnology, 139–140
  predictive, 79–80
  preimplantation genetic diagnosis, 72
  symptomatic persons, 81
  vs. screening, 71
Diamond v. Chakrabarty, 12
Discrimination, 84, 85–86, 141, 143
Diseases and disorders
  Alzheimer's disease, 54–57, 55t, 56f, 140
  asthma, 43–45
  autosomal recessive disorder, 73f
  bacteria, 137
  cancer, 57–60, 58f, 59(f5.10), 59(f5.11), 59(f5.12)
  Center for Inherited Disease Research, 92
  chromosomal disorders, 52
  common genetically inherited diseases, 53–54
  cystic fibrosis, 52(f5.1), 60–62, 60f, 61f, 89
  diabetes, 62–63, 63f, 63t
  environmental factors, 43, 50, 50f
  gene therapy, 134–137, 135f, 136f
  genetic susceptibility, 45
  hemochromatosis, 53(f5.3)
  Huntington's disease, 31, 63–65, 64f, 88, 134
  "inborn errors of metabolism," 5–6
  leading causes of death, 54t
  molecular medicine, 10–11, 101–102
  mtDNA linked, 27
  multifactorial disorders, 51–52
  muscular dystrophy, 65–67, 66f
  phenylketonuria (PKU), 67, 68(f5.20)
  preclinical disease detection, 101, 133

Nucleotides, 8, 15
Numerical chromosomal aberrations, 41

O

Oligonucleotide hybridization, 78–79
On the Origin of Species by Means of Natural Selection (Darwin), 2
Oogenesis, 22, 26(2.17)
Oregon Regional Primate Center, 108
Organ transplantation, 151
Organic chemical compounds, 103
Organismal cloning. *See* Cloning, reproductive

P

Painter, Theophilus Shickel, 8
Panel on Scientific and Medical Aspects of Human Cloning, 113
Paralysis, 110
Parkinson's disease, 109, 110, 111
Parthenogenesis, 111
Particle-mediated transformation, 122
Particulate theories of heredity, 1
Patents
    gene sequences, 91, 92
    genes and DNA sequences, 13
    genetically modified crops, 124
    genetically modified organisms, 12
    terminator technology, 124–125
Paternity, 84
Pauling, Linus C., 8–9
PCR. *See* Polymerase chain reaction (PCR)
Pea plant experiments, 3–5, 5f, 6f
Peptide bonds, 17
Performance-enhancing genes, 137
Perkin-Elmer Corporation, 92
Perlegen Sciences, Inc., 103
Personality traits, 49
Pest resistance, 124
Pesticides, 125–126
Pharmaceuticals
    Alzheimer's disease, 57
    anticancer drug resistance, 134
    genetically engineered drugs, 12
    molecular farming, 133
    nanotechnology delivery, 139
Pharmacology
    genetic tests, 81
Phenotypes, 24–25, 27f
Phenylketonuria (PKU), 52, 52(f5.2), 67, 68(f5.20)
Pigs, 108
PKU (phenylketonuria), 52, 52(f5.2), 67, 68(f5.20)
Plant genomics, 122
Plants
    genome sequencing, 13
    Mendelian genetics, 3–5, 5f, 6f
Plasmids, 105, 106(f8.1), 106(f8.3)

Plomin, Robert, 48
Pneumonia, 8, 8f
Point mutations, 41
Pollack, Andrew, 136, 137
Polygenic inheritance, 44, 45, 51–52
Polymerase chain reaction (PCR)
    cloning, 106–107
    genetic prenatal testing, 75, 77f
    invention of, 12
    Mullis, Kary B., 88
    steps, 89(f7.3)
Population screening, 82–83
Populations
    evolution, 33
    gender ratio, 28–30, 31t
    growth, 34
    Hardy-Weinberg Equilibrium, 7
Potatoes, 132
Preclinical disease detection, 101, 133
Predictive testing, 65, 79–80
Preformationist theories of heredity, 1
Preimplantation genetic diagnosis, 72, 74(f6.3)
Prenatal testing
    laboratory techniques, 75–79, 75t, 77f, 78f, 79f, 80f, 81f, 82f
    overview, 72–75
    public opinion, 147t, 148f
Prescription drugs. *See* Pharmaceuticals
President's Council on Bioethics, 113–114
Primates, 108
Private *vs.* public genome technology initiatives, 92–93
Probability, genetic, 25–26
Progenitor cells, 112
Prokaryotic cells, 21
Proposition 71 (California), 116–117
Prostate cancer, 46
Proteins
    amino acids, 17
    DNA codes, 15–17
    genetic synthesis, 18–20
    protein synthesis, 10
    structures, 17(f2.3), 18(f2.5)
Psychological issues, 84–85, 86
PubChem, 103
Public opinion
    cloning, 119f, 143–144, 144t
    evolution, 36–39, 36f, 37f, 38f, 38t
    genetic information, 141–142
    genetic technologies awareness, 144f, 145t
    genetic technology as next step in human evolution, 148t
    genetic testing, social implications of, 149f
    genetic testing awareness, 146t
    genetically modified crops, 129, 129f, 130f
    prenatal genetic testing, 147t, 148f

reproductive genetics, 150f
    stem cell research and cloning, 117, 117f, 118f
Public Patent Foundation, 125
Public policy
    genetic testing and research, 143
    genetically modified crops, 125–126, 129
    human cloning, 112–113, 114–117
Public schools
    evolution *vs.* creationism, 38–39, 38t
Puffer fish, 95–96
Punctuated equilibrium, theory of, 42
Punnett squares, 25–26, 28(f2.20)

Q

Quammen, David, 36

R

Rabbits, 108
Race/ethnicity
    The Bell Curve: Intelligence and Class Structure in American Life (Herrnstein and Murray), 47
    cystic fibrosis, 60
    genetic discrimination, 141
    life expectancy, 32t
    sickle-cell disease, 68–69
    social Darwinism, 35
    Tay-Sachs disease, 70
Radiation, 42
Raelians, 109
Reagan, Nancy, 114
Reagan, Ronald, 114
"Rebellious Teens? Genetic and Environmental Influences on the Social Attitudes of Adolescents" (Abrahamson, Baker and Caspi), 49–50
Rebelliousness, 49
Recessive inheritance, 24–25, 28(f2.19), 29f
Recloning, 108
Recombinant DNA, 12, 87, 105–106
Regeneration of body parts, 111
Regulation
    biotechnology, 125–126, 129
    gene therapy, 136
    genetic testing, 84
    reproductive genetics, 150(f10.5)
Reliability, test, 71–72
Religion, 35, 36–37, 49–50
Replication
    bacteria, 11
    DNA, 16f
    meiosis, 22, 25f
Reproductive genetics
    cloning, 107–109, 107f
    ethics, 113–114
    growth and reproduction, 21–24
    humans, 109